Engineers as Leaders

Engineers as Leaders

Beyond Logic!

JESSE L CALLOWAY, Ph.D.

Copyright © 2019 by JESSE L CALLOWAY, Ph.D.

Library of Congress Control Number:		2019906850
ISBN:	Hardcover	978-1-7960-3720-3
	Softcover	978-1-7960-3719-7
	eBook	978-1-7960-3718-0

All rights reserved. No part of this book may be reproduced or transmitted in any form or by any means, electronic or mechanical, including photocopying, recording, or by any information storage and retrieval system, without permission in writing from the copyright owner.

The views expressed in this work are solely those of the author and do not necessarily reflect the views of the publisher, and the publisher hereby disclaims any responsibility for them.

Any people depicted in stock imagery provided by Getty Images are models, and such images are being used for illustrative purposes only.
Certain stock imagery © Getty Images.

Print information available on the last page.

Rev. date: 07/10/2019

To order additional copies of this book, contact:
Xlibris
1-888-795-4274
www.Xlibris.com
Orders@Xlibris.com

CONTENTS

Preface vii
Acknowledgments ix

Chapter 1: So You Want to Lead? 1
Chapter 2: Leadership Defined 6
    2.1: Theoretical Leadership Models 10
    2.2: Leadership Levers 13
    2.3: Emotional Intelligence and Leadership 21
    2.4: Resonant Leadership 25
    2.5: Transformational Leadership and Credibility 28
    2.6: The Role of Psychology in Leadership 32
    2.7: Predicting Leadership Behavior 37
Chapter 3: Power and Leadership 40
Chapter 4: Critical Research Parameters 54
    4.1: Problem 55
    4.2: Method and Procedure 56
Chapter 5: Introduction to the Research 61
    5.1: Literature Review 62
    5.2: Literature Review Detail 63
    5.3: Literature Review—Beyond the Gap 64
Chapter 6: Methodology 68
    6.1: Literature Review-Based Research Paradigm 69
    6.2: Measuring Instrument 70
    6.3: Measuring Instrument Validity 72
    6.4: Measurement Instrument Reliability 73
Chapter 7: Research Execution 75
    7.1: Population and Demographics 76
    7.2: Data Analysis—Engineering Education
    and (TL, XL, PA) 79

	7.3: Data Analysis—TL Constituents	85
	7.4: Data Analysis—Predominant Leadership Style	90
	7.5: Data Analysis—Percentile Comparisons	95
Chapter 8:	Conclusions and Relevance	104
	8.1: Generalizability	108
	8.2: Limitations and Areas for Future Research	109
	8.3: Where Do We Go from Here?	110
Chapter 9:	Meet Josef Brilliant	111
	9.1: Communication—It's More Than Talking	116
	9.2: Effective and Efficient Communications	121
	9.3: Managing Multiple Priorities Concurrently	126
	9.4: Josef's Expert Power in Action	137
Chapter 10:	Josef's Big Break	140
Chapter 11:	Leading from Behind	150
Chapter 12:	Introduction to the 5Ps for Exhibiting Presentation Leadership	160
Chapter 13:	Bringing It All Together	182

References	187
Appendix A: Detailed Literature Gap Analysis	197
Appendix B: Multifactor Leadership Questionnaire	199
Appendix C: Sample MLQ5X Scoring Key	201
Appendix D: Normative Tables	203
Appendix E: Full-Range Leadership Theory	205
Appendix F: Survey Process	209
Appendix G: Nonparametric Testing	211
Index	217

PREFACE

AT A RELATIVELY young age, I retired from the corporate sector (where I served in multiple executive levels) and went on to complete a PhD in engineering management, successfully defending my leadership-oriented dissertation and graduating in May 2015. I am currently president of my company, Leadership and Motivation Consultants, a certified executive coach, adjunct faculty member at Old Dominion University in Norfolk Virginia, member of a board of governors for higher education, prior member of the NCA&T Engineering Advisory Board, and recipient of the prestigious TMCF Corporate Leadership Award.

I once had an instructor who said that if you did not try to make things work that were either broken or not intended to work in the manner that you desired, then you were probably not true engineering material. Indeed, there are many stereotypes regarding engineers and their inability to do anything except design new gadgets, solve technical problems, and create highly sophisticated software solutions.

The early stages of my career fell in line with those stereotypes. Beginning my career as an electrical technician, I always had a penchant for electronics and trying to design that newfangled product that I would patent, sell, and leverage for my retirement. My first attempt in this area came when I designed a gadget (electrical/electronic and mechanical) to start an unoccupied vehicle in the winter or summer months to allow for the warming or cooling of the vehicle to a desired state prior to the driver or passenger entering. It was at this point that I learned that the cost for a simple "patent pending" stamp was around $800.00. In the 1980s, this was a lot of money for a young man from modest means. Following this and other designs, including proportional-integral-derivative (PID) and programmable logic controllers (PLCs), I subsequently realized that my aspirations for achieving the American dream were not going to be financed by patented products. Nor would I be able to get there working behind a desk cluttered with lots of drawings, sketches, and printed circuit cards.

For me, the path had to take a new direction. I had to leverage my technical capabilities and act in a less technical way. For me, the answer was leading people who may have little to no knowledge of the technical world as most engineers know it.

As I continued to climb the corporate ladder, I also realized that there were many other engineering-oriented individuals who, like me, were capable of stepping beyond the technical realm and into the world of leadership. Indeed, it was this thinking that served as the basis for my design and implementation of an Engineering Leadership Development Program (while in the corporate sector) for the sole purpose of building leadership capability in engineers.

The aim of this book is not at all to suggest that engineers should vacate the holy grail of engineering and aspire to become leaders. Indeed, I could not imagine a world without the tremendous contribution of practicing engineers. However, for those who long for something else, I implore you to read on!

ACKNOWLEDGMENTS

IN HIS POEM "Ulysses," Alfred, Lord Tennyson wrote, "I am a part of all that I have met." Likewise, this book would not be possible were it not for the tremendous contributions of my family, friends, and colleagues. The unbiased feedback offered by my wonderful daughter and the patience exhibited by my wife of more than thirty years deserve special recognition.

In loving memory of my family members who are reading this book from the heavens, I am donating 20 percent of all proceeds from sales of this book to charitable organizations, including dementia and cancer research centers.

CHAPTER 1

So You Want to Lead?

CONGRATULATIONS ON YOUR desire to move up the economic food chain. You are in good company as reflected by the November/December 2018 Issue of a *Harvard Business Review* (HBR) article titled "The Best-Performing CEOs in the World 2018." According to HBR, "Thirty-four of the top 100 CEO's possessed an engineering degree, while only 32 had MBA's." As an engineer, your fascination with measuring things has probably piqued with this HBR article, and you may be wondering just how they determined that these CEOs were the best-performing. They actually used very relevant metrics—total shareholder return (including dividends reinvested and adjusted for country and industry) as well as market capitalization. I realize that these terms are not typical of that offered by engineering schools, and you no doubt have the desire to learn more in this area and I encourage you to do so. Buttressing this view, "The future continued success of U.S. businesses and industry rests upon having more practicing engineers with knowledge in topics such as: Principles of Leadership and Management" (Compton-Young, et al., 2010). At this point, you may be asking why the engineering degree is so valuable in leading a business. According to an earlier HBR publication, "Studying engineering gives someone a practical, pragmatic orientation … engineering is about what works, and it breeds in you an ethos of building things that work—whether it's a machine or a structure or an organization" (November 2014). Said differently, engineers by training are great problem solvers. And there is no more evident an opportunity to apply such skill than leading a business thorough its most turbulent times—fierce competition for a shrinking customer base, the incessant goal of identifying and fulfilling an unmet consumer need, business globalization, or developing the most efficient approach for providing goods and services. *Forbes* offered perspective on the topic as well in their May 2014 article titled "Why Engineers Make Great CEO's," which stated, "Engineering has long been ranked as the most common undergraduate degree among Fortune 500 CEO's." So far so good, right? Well, there is a slight catch as also reflected in the same *Forbes* article, which states, "Many [engineers] lack emotional intelligence and the necessary leadership, people management, and communication abilities—soft skills which can be addressed by training to assist their transition into the management arena." The operative point of this statement is that these "soft skills" are only a few steps away and building capability in this area begins with what you are doing at this very moment, namely reading this book.

So let's start here. I must admit that for many of my younger years, I thought *engineer* referred to someone steering a train. Don't chuckle too much; this view has been shared by many others for years as pointed out by Weingardt (1994), "A recent Gallup poll indicates that one-third or American citizens think engineers drive trains." Indeed, as a child, I enjoyed electric train sets and dreamed of one day possibly becoming an engineer. Little did I know at the time that, although I would never drive a train, I would serve in the role of electrical engineer but also move on to serve as vice president of process commercialization and engineering. And I would subsequently take those same leadership and problem-solving skills with me to the position of vice president and general manager of processing and manufacturing. While both positions fully leveraged some of my undergraduate learning, I also recognized the importance of having skills and experiences that were not offered as prerequisite engineering curriculum courses. Often referred to as people skills or soft skills, which were discussed previously, these skills nicely complemented the technical and problem-solving skills. This view of the role that such complementary skills play in engineering leadership is broadly recognized. Indeed, "An engineer is hired for her or his technical skills, fired for poor people skills, and promoted for leadership and management skills" (Yao and Russel, 1997). The value of engineers as leaders is well known and is not limited to highly technical fields. Amazon CEO Jeff Bezos, for example, majored in electrical engineering and computer science. Rex Tillerson, the former CEO of Exxon Mobile (Secretary of State), majored in civil engineering. General Motors CEO Mary Barra majored in electrical engineering. Larry Page, the cofounder and CEO of Google, majored in computer engineering. Branching out a bit, according to a study conducted by a consulting firm in 2011, McDonald's Corp. and Bank of America were then led by engineers. And of course, Tim Cook, the CEO of Apple, majored in industrial engineering.

The path from working as an engineer to serving in leadership is not necessarily cookie-cutter for everyone. For me, the greatest challenge in moving from engineer to leader was realizing that, unlike an electronic circuit, resolving problems and opportunities associated with humans and organizations was much more complicated than designing an engineering solution to a technical problem or simply extracting one integrated circuit and inserting a new one. It involved recognizing that decisions were not all binary or black-and-white choices. Instead it was about getting to the lighter or darker shades of gray. A further elaboration of this point is illustrated by figure 1-1.

Figure 1-1: Transistor Switching Circuit

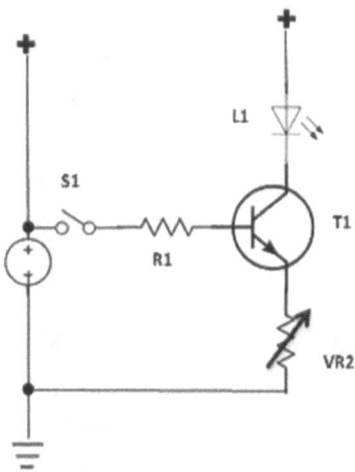

The basic switching circuit is enabled (turned on) when switch S1 is closed, thereby supplying a positive voltage (and current) to the base of transistor T1 through current-limiting resistor R1. The light emitting diode (LED or L1) is then illuminated as current flows through the transistor collector and emitter and then through variable resistor (VR2) to ground. If the LED is not as bright as desired, VR2 may be varied (reducing resistance and thus increasing current flow), which will, in turn, increase the brilliance of the L1. Conversely, VR2 may be adjusted in the opposite direction (increasing resistance) to reduce L1 brilliance.

This circuit is not only very simple but also very logical. Unfortunately, leading people does not always follow a logical pattern. For example, if the resistor is not doing its job, it can be removed with no questions asked, and another one can be inserted in its place. If an employee is not very motivated to execute a goal or task, he or she is not so easily replaced. Indeed, removing an employee from a job is very often a lengthy—possibly even litigious—process, not to mention the potential emotional and economic impact that doing so might have on the employee, his or her family, and the business. And what recourse is there if the replacement employee also does not perform? So in leadership, the job becomes more than simple logic. It becomes people-oriented. And in this particular example, the goal is to not simply terminate the employee but rather to motivate that same employee to perform his or her job consistent with organizational needs. Some of you may think that the way to get an employee to perform at a high level is to implement coercive methods such as withholding pay or other perceived rewards. Here's a news flash. This very transactional approach, as we will discuss at length in subsequent chapters, is not the answer. But don't worry. We will also discuss at length leadership styles grounded in motivation and organizational transformation.

The focus of this book is to peel back the layers of the onion of leadership that are geared specifically toward engineers. Bonasso (2001), put it this way, "Two significant boxes create the engineer's world. The first box is created by the narrowly focused technical education process. The second and most confining box is defined by the world of technical work. Engineers and scientists are gifted problem solvers ... breaking out of these two boxes and applying their problem-solving skills and talents more broadly can offer a larger role in societal leadership." Throughout this book, the message will be consistent. Engineering technical skills will take you only so far. Moving up the economic food chain requires an expansion of these technical skills. Brazil and Farr (2007) certainly concur with the value added by augmenting such skills. They state that "most senior engineers are successful because they have demonstrated technical excellence and some management ability ... to make the transition to respected leader; they must continue to develop ... qualities of a leader." Okay, now that we have identified an ostensible gap when contrasting typical engineering skills with the same required for successful leadership, it would be appropriate to delve deeper into the notion of leadership. What exactly is leadership?

CHAPTER 2

Leadership Defined

LET'S NOW DISCUSS the notion of leadership, which has been described and defined at length in textbooks and literature of all types. The leadership definitions and characterizations that follow are peppered with some specificity to engineers. According to Bonasso (2001), "Leaders are people who act as guides for groups of people bringing them together to unleash their potential." In discussing a course on engineering leadership, McCuen (1999) wrote, "Leadership consists of the knowledge and skills that the individual possesses and employs to persuade others to enthusiastically work toward the completion of the plan of action that the leader has established ... A leader needs to effectively motivate subordinates to successfully achieve the established goals ... leadership is more than just the application of managerial skills." This definition, in essence, speaks to fundamental human motivation factors. Another descriptor is provided by Katz and Kahn, as cited by Johns and Moser (1989), characterizing leadership as "any act of influence on a matter of organizational relevance." This would imply that one of the principle responsibilities of a leader is to move an organization forward in a manner consistent with the best interests of the business. This notion of consistency should not be understated. All too often instances have been documented of leadership application in a manner inconsistent with an organization's stated relevance as happened with Enron and Global Crossings in the 1980's and early 2000's respectively. According to Crumpton-Young and others (2010), "engineering Leadership is the ability to lead a group of engineers and technical personnel responsible for creating, designing, developing, implementing, and evaluating products, systems, or services." Although this perspective circumscribes leadership opportunities for engineers, it also appears to highlight the engineers' resourcefulness.

Burns (1979), as also cited by Johns and Moser, said, "I define leadership as leaders inducing followers to act for certain goals that represent the values and the motivations of both leaders and followers." This descriptor implies that there exists a sort of connectedness between and among the leaders and followers. Perhaps this is the intent of the following comments offered by Boyatzis and others (2005): "Great leaders are awake, aware, and attuned to themselves, to others and to the world around them."

Chemers (2001), as cited by Kark and Yaffe (2011), defines leadership as "a process of social influence through which an individual enlists and mobilizes the aid of others in the attainment of a collective goal." This perspective, similar to the preceding comments, suggests that leadership involves not only the leader but also the stalwart participation of those to whom the leader is looked to for guidance and direction. A key distinction here is the reference to "social influence," which could suggest that the leader's actions and behaviors are somehow swayed by those with whom the leader interacts. Another key distinction in this leadership characterization is reference to "a collective goal." Both social and collective could imply the leader's accountability to his or her associates to whom he or she is entrusted to lead.

According to Shaw, as cited by Hartman and Jahren (2015), "Engineering leadership is the process of envisioning, designing, developing, and supporting new products and services to a set of requirements, within budget, and to a schedule with acceptable levels of risk to support the strategic objectives of an organization." Similar to comments offered by Compton-Young and others, Shaw's perspective fails to recognize the capacity of engineers to flourish as leaders beyond the stereotypical engineering paradigm.

According to Northouse (2013), leadership "is a process whereby an individual influences a group of individuals to achieve a common goal." This definition attempts to transform leadership from the psychological realm to a sort of interdependent network. *Merriam-Webster (2007)* defines the word process as follows: "a forward or onward movement ...

something going on ... a natural phenomenon marked by gradual changes that lead toward a particular result." With this definition in mind, it may be inferred that leadership, viewed as a process, is about a leader providing direction or inputs to the followers who, in turn, move forward with that direction (transforming it into a qualitative or quantitative deliverable), seeking consistency of the output with the initially provided direction (input). Systematizing this thinking (see figure 2-1) would yield a codependence characterized by inputs, processing, and outputs with the addition of a feedback mechanism to close the loop.

Figure 2-1: Leadership Systemization

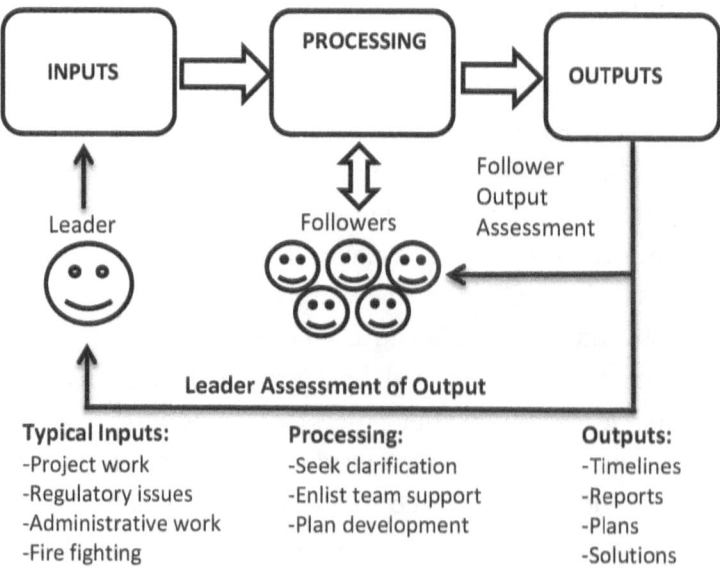

Focusing on figure 2-1, you can see the leader provides direction in the form of inputs to the followers who in turn process that direction to develop an output. The initial output is reviewed by followers and modified to fit their interpretation of leader inputs. Follower output is submitted to the leader for assessment. Upon assessment by the leader, the output is either reworked by the followers, or it is accepted by the leader, who then provides a new input for processing. (The topic of inputs, processing, and outputs will be discussed further in chapter 10.)

Hesburgh (1971), as cited by Johns and Moser (1989), "gave an inspiring definition of leadership, 'the mystic of leadership, be it educations, political, religious, commercial or whatever, is next to impossible to describe, but wherever it exists, morale flourishes, people pull together toward common goals, spirits soar, order is maintained, not as an end in itself, but as a means to move forward together it requires courage as well as wisdom.'" This leadership characterization suggests that almost in a supernatural manner, a leader is able to galvanize the team, the organization, and perhaps a nation. Further, those who receive the message all march with one accord en route to accomplishing the unimaginable. Nowhere was this perspective more evident than in the following passage provided by Burton (2009), quoting then President Kennedy: "We choose to go to the moon in this decade and do the other things … not because they are easy, but because they are hard." Such direction provided in the twenty-first century would not be unexpected and may seem quite trivial. However, this proclamation was issued at a time when those accountable for the ultimate deliverable were asking fundamental questions regarding how to make it happen. This perspective was evidenced by the comments written by Burton (2009), which state, "Immediately following Kennedy's announcement, NASA managers asked themselves, 'how do you get to the moon.'"

Despite having touched on several leadership perspectives, it was not my intent in the foregoing discussion to present exhaustive commentary in this area. Indeed, as written by Burns (1978) and cited by Johns and Moser (1989), "The list of well-reasoned definitions of leadership could go on and on."

With these comments, discussion now shifts toward reviewing attempts to prototype leader functions and behaviors.

2.1: Theoretical Leadership Models

Modeling leadership approaches and patterns has proven quite useful, particularly in an academic setting where the students have yet to experience firsthand the joys and pains of leadership. Among the many leadership models is *contingency theory*, which is, according to Fiedler and Garcia (1987) and cited by Northouse (2013), "the most widely recognized [contingency theory model]." As the name implies, the model posits that leadership styles and responses are contingent on various situations and based on these situations, characterizes the leader as either "relationship motivated" or "task motivated." Specific situations may be characterized in terms of "leader-member relations, task structure, and position power."

As an example regarding leader-member relations in an environment where trust and good overall perception of the leader is experienced, such relations are "defined as good." Task structure refers to "the degree to which requirements of a task are clear and spelled out," while position power has to do with "the amount of authority a leader has to reward or to punish followers [both of which, to some degree, parallel transactional leadership, which we will discuss later]" (Northouse 2013).

From these comments, it should be clear that the contingency theory model suggests certain paths be embarked on, paths driven by the situation at hand. For example, if the situation to be addressed is "moderately favorable or moderately unfavorable," the model suggested approach is one that is "relationship oriented." Further, "if a leader is moderately liked and possesses some power" under somewhat ambiguous job conditions for subordinates, a "relationship orientation" should provide the best chance for success.

The underlying premise for this contingency approach is that leaders must be perceptive enough to recognize certain situations and circumstances that, in turn, will prompt them to adjust that environment to better match their leadership approach. Another way to interpret this is as follows: "When leaders can recognize the situations in which they are most successful, they can then begin to modify their own situations" (Ivancevich and Matteson 1993). This approach may seem a bit counterintuitive as it suggests that rather than exhibit leadership flexibility, the leader should modify the situation to one more compatible with personal style. The premise for this approach was that "Fiedler [was] not particularly optimistic that leaders [could] be retrained successfully to change their preferred leadership style" (Ivancevich and Matteson 1993).

Path-goal theory, expanding upon the contingency theory approach, "suggests that a leader must adapt to the development level of subordinates [emphasizing] the relationship between the leader's style and the characteristics of the subordinates and work setting" (Northouse 2013). An important point to be made regarding path-goal theory is that it is based on subordinate perceptions of their leaders' work and how they achieve goals within their particular work environment. Similar to "other situational or contingency leadership approaches, the path-goal attempts to predict leadership effectiveness in different [leadership] situations" (Ivancevich and Matteson 1993).

House and Mitchell (1974), as cited by Northouse (2013), offer four leadership behaviors applicable to the path-goal theory, which include "directive, supportive, participative, and achievement-oriented." As the stated behaviors would suggest, the directive style focuses more on providing direction, whereas the supportive and participative approaches tend to enlist collaboration from subordinates while achievement orientation seeks to build capability among subordinates. The key takeaway from this approach is that the leader must be fully aware of the capability of his or her subordinate staff as well as their motivational needs and the overall work environment. It is only after such analysis that the leader will be positioned to apply the appropriate leadership style.

Although it is not my intent here to address all possible theoretical leadership models, two remaining theoretical models have garnered quite a bit of support, namely "leader-member exchange theory" and the "Vroom-Jago model of leadership." Leader-member exchange (LMX) is predicated not simply on the style of the leader or subordinates or even the specific situation at hand. Instead it "takes still another approach and conceptualizes the leadership as a process that is centered on the interactions between leaders and followers" (Northouse 2012).

To this point, we have assumed a degree of universality among subordinates. Leader-member exchange by design seeks to segregate subordinates into two distinct groups—the "in-group" or the "out-group," both of which are contingent on the leader to subordinate relationships (Northouse 2012). Generally speaking, the more collaborative and ambitious the subordinate, the greater the likelihood that this person will be aligned with the in-group. Those falling outside this area would obviously be more closely aligned with the out-group. And depending on the group allocation, the leader-subordinate interaction varies accordingly. For example, as noted by Dansereau and others (1975) and as cited by Northouse (2012), "Subordinates in the in-group receive more information, influence, confidence, and concern from their leaders than do out-group subordinates."

Barge and Schlueter (1991) suggested that LMX theory is predicated on the notion that "in-group relationships will be associated with higher levels of employee satisfaction and productivity." The effectiveness of LMX has been empirically confirmed as "in-group relationships are not only positively associated with increased employee satisfaction (Ferris 1985; Graen and Ginsburgh 1977) but with employee performance as well (Liden and Graen 1980; Tjosvold 1984; Vecchio 1982)" as also cited by Barge and Schlueter (1991).

This perspective of employee effectiveness is shared by Dubrin (2010), who wrote based on study results, "The quality of the relationship with the leader had an impact on the effectiveness of influence tactics … a poor relationship with the leader resulted in less [coworker to coworker assistance while] a positive relationship with the leader positively related to helping behavior." While LMX can result in positive contributions by certain team members, there are some negative implications as well. This is principally due to the variations in business relationships. Indeed, Dockery and Steiner (1990), as cited by Suleyman (2011), stated that "in high-quality interactions, leaders establish closer relations with only a few key subordinates, the (in-group) due to limited resources [consequently] they provide (in-group) members with support and resources beyond the employment contract."

The Vroom-Jago model has to do with decision making and the degree to which subordinate involvement should be taken into account when making such decisions. The fundamental assumption for use of this model was that "no single leadership style was appropriate" and that leaders must exhibit flexibility even if doing so requires the leader to modify his or her style to fit the situation at hand. By design, the model also considers the types of decisions with which leaders are faced, namely "individual and group." As implied by the terms individual and group, the former decisions have to do with leader decisions that affect only one member of the team while the latter addresses decisions that "affect several followers." Because of the complexity associated with use of this model (driven by the highly variable nature of decisions to be made), "decision making heuristics, or rules of thumb, have been developed" (Ivancevich and Matteson 1993).

As previously mentioned, leadership models aimed at improving leader effectiveness, be it through subordinate motivation, performance management, decision efficiency, or otherwise, abound in the related literature. While it was not my intent in this section to comprehensively address any and all such theoretical approaches, those mentioned should adequately introduce the topic and possibly precipitate additional enquiry, which is left to you to pursue at will.

2.2: Leadership Levers

Thus far our discussion of leadership has considered leadership characterizations, theories, and models for the most part. Yet there are other mechanisms at the leader's disposal that may also be of assistance in moving the needle of organizational effectiveness. Three such critical tools are leveraging teams and their associated infrastructure, receiving and delivering feedback, and leadership coaching.

Teams

Regarding teams, there is much to be said of the progress that can be made working in a collaborative group versus flying solo. Strength in numbers is perhaps nowhere more evidenced than by nature's wolves. Wolves, like humans, achieve even the most critical and fundamental goals (such as hunting) in packs or groups. Operating in a sort of hierarchy, wolves are pack animals that communicate by the gestures of their heads, bodies, and limbs, thus maintaining order in the pack. In the same way that a father wolf obtains food for the family, the leader of a human team removes barriers to effectiveness and quenches the members' hunger for challenge (Funk and Wagnalls 1995).

One way to view a team of people interacting interdependently is to consider them in a systems context. Doing so is consistent with the perspective described by Kets and others (2007), who suggest that "a system is a set of interacting units with relationships among them." Human teams, particularly when operating at optimum efficiency, provide such a relationship. Teams may be formed formally (such as when designated by a leader or sponsor) or informally, whereby a group of employees with a common goal recognizes the benefits of operating collectively.

Despite the motivation for group formation, formal and specific stages of behavior occur before optimization. Bateman and Zeithaml (1993) describe these stages as "forming, storming, norming and performing [then] adjourning." Although each stage is relatively self-explanatory, storming is perhaps the most controversial aspect of group and team development. It's where each member seeks to define him or herself and lobby for respective contributions to the team. While conflict may be minimal when each team member is perceived as bringing different yet valued skill sets to the team, such is not necessarily the case when one or more members appear to exhibit expertise in the same area. It is this situation that may give rise to conflict over such matters as how to decide which team member is most suitable for a needed role.

The team leader, formal or informal, is most often looked to for deciding and allocating team member roles and contributions. The leader is also accountable to ensure a clear understanding of team potency, which is, according to Champion and others (1993) and as cited by Hu and Liden (2011), "team members' shared beliefs about their collective capabilities." This is a critical aspect of team evolution and effectiveness. If team resource capabilities are either underutilized or overstated, the result will be team suboptimization. So the team must trust that the leader is best positioned to make such determinations while concurrently fully making use of and valuing whatever a team member has to offer. This perspective is supported by Lam, Peng, and Schaubroeck (2011) in the following statement: "Members' trust in their leaders is critical for effective team performance and potency."

It is also important to note the leaders' influence on the team given certain cultural and social settings. Earlier we discussed collectivism and the role that such approach might play in leader-team dynamics. Further, according to Yukl (2010) and as cited by Chang, Johnson, Mao (2012), and Venus, "Leadership is a social process of exerting influence over the thoughts, feelings, and actions of others." With this in mind, it was also noted in the same report that "leaders' group based identities have also been found to spill over to their followers." Thus, collective consideration should, in turn, promote healthy member collaboration.

Although providing team leadership, recognizing the unique contributions of each member, and appropriately allocating resources among team members allow a team to progress toward optimization, empowerment can accelerate such optimization. Indeed, according to a study implemented by Leach (1998), as cited by Clegg, Cordery (2002), and Wall, which also considered enhanced feedback, "System performance improved considerably following empowerment." The essential goal of empowerment is for the leader to provide the team with needed resources and then get out of the way and serve the team. Said differently, empowerment is simply "a means of granting work-related decision-making authority to employees as a means of enhancing performance" (Menon, 2001). Through empowerment, team members derive a sense of ownership for goals and accomplishments. In this way, they execute with passion and quality.

While charged with providing direction to the team and steering them along the correct courses of action, the leader must nonetheless be careful about advocacy of certain actions. Granted, the team must take certain paths, and the de facto decision maker in such situations is the team leader. However, when opportunities exist to engage the team in making decisions or plotting future courses of action, there is a somewhat tacit expectation among the team that the members are expected to participate. The astute team leader recognizes these opportunities as well as the possibility of a disengaged team if such opportunities are not appropriately pursued or if team members perceive their input to be of little value. Indeed, according to Vroom (1997), "Strong advocacy by the leader of a particular course of action along with critical judgments of alternatives proposed by others, might reasonably be expected to decrease [team] participation." More often than not, the team's success or failure rests with the team leader. Consequently, the team leader's strengths and development opportunities are often ultimately measured by overall team performance.

Feedback

A frequently used tool aimed at honing a leader's effectiveness is the 360-degree feedback instrument. Although the emphasis of this section, for the purposes of reviewing 360-degree feedback, has to do with individual improvement, it may also be used "for succession planning, merit raises, performance appraisals, and downsizing" (Capritella 2002, as cited by Crispo and Sysinger 2012). Throughout this writing, reference has been made to leader-to-group interactions and their importance in the leadership arsenal. Here, we delve a bit deeper into a formal feedback instrument, the 360-degree form, as well as review aspects of its supporting infrastructure.

We begin with an efficient definition of the instrument as follows: "The 360 degree feedback is a questionnaire that is completed by the participant, participant's supervisors, coworkers, peers, and subordinates" (Crispo & Sysinger 2012). Hence, reference to the tool as being "360 degree" feedback adequately articulates the extent to which organizational feedback (participants' strengths and weaknesses) is provided. In fact, in some instances, the word *weaknesses* is often supplanted with "development opportunities" to assure the highest chance for success in the leader's acceptance of such feedback. (The latter descriptor may be viewed as less critical.) This is an important property of the 360-degree process as the intent is to receive balanced feedback from the organizational levels with which the participant most frequently interacts. This perspective is shared by Hellervik, Hazucha, and Schneider (1992), as cited by Carless, Mann, and Wearing (1998) as follows: "Obtaining information on an individual's performance from multiple sources enhances the credibility of the information and therefore, presumably the individual's motivation to change his or her behavior."

Aimed at providing a more coherent review of the 360-degree feedback questionnaire composition, we will focus on the "global executive leadership inventory (GELI)" described by Kets and others (2007), which includes the following key components for inventory: "visioning, empowering, energizing, designing and aligning, rewarding and giving feedback, team-building, outside stakeholder orientation, global mindset, tenacity, emotional intelligence, life balance [and] resilience to stress." The significance of the balanced approach coupled with use of the feedback circle cannot be overstated and offers the best chance for success in elevating behavior to the desired state.

Those in the feedback circle (the raters) assess each of these components and provide information to the one being rated about how well that person scores within the component (for example, in team-building) and also how those scores compare to an average score. This feedback helps leaders understand better the emotional lives of the ones being rated.

The feedback might be supplemented with a personality audit. It may focus on a person's motivation and emotional management (such as trustful versus vigilant or extroverted versus introverted). Archetypes feedback reveals how the one being rated deals with people and situations (such as whether he or she is a strategist or a coach) (Kets et al. 2007). In the end, the comprehensive nature of this feedback is expected to fully convey those personality characteristics that, if modified, would facilitate increased individual and organizational effectiveness. (Beginning in chapter 5, reference will be made to the Multifactor Leadership Questionnaire [MLQ]. Similar to the GELI, it may be used for receiving feedback on a 360-degree basis.)

Coaching

The final leadership tool I wish to discuss in this section is executive coaching. How often have we seen the underdog team miraculously execute an amazing come-from-behind victory or heard about an impoverished elementary school that succeeded against all odds in meeting testing score requirements? These stories convey the essence of the power behind coaching. Yet effective coaching is not confined to circumstances offering low probability for success. Today many executives depend on coaching, either formally or informally, to buttress their success. Indeed, according to the Chartered Institute of Personnel and Development (2010), as cited by Baban and Ratiu (2012), "Two-thirds of organizations report using coaching."

You might wonder how effective executive coaching can be, given that managers are seasoned and to some degree, unyielding. A metaphorical adage says, "You can't teach old dog new tricks." Nonetheless, retraining seasoned executives by coaching them is quite doable and beneficial. According to Kets and others (2007), "People whose personality characteristics have been largely formed (this includes most people over 30) can still make significant changes in their behavior." While coaching strategies vary depending on a client's operating environment, this book assumes the typical corporate or business environment with a leader who has no lineage ties within that organization.

Although a great deal of coaching efficacy may depend on the one being coached, Nelson and Hogan (2009), as cited by Baban and Ratiu (2012), stated that "coaching in general can be a more productive and impactful process if coaches engage in a well-planned and intentional manner." One of the first steps used in executive coaching is to provide the client with unbiased feedback such as that offered via the 360-degree instruments discussed previously.

Such an approach is extremely valuable given that effective coaching has to be as objective as possible—a sort of reality check. According to De Berg and others (2012), "Feedback is particularly relevant in coaching practices where it is provided to support self-awareness, learning, and to improve performance." Indeed, many executives have a differing perspective of their interactions, and given their positions of power, they often face little resistance regarding their beliefs about themselves. This view is shared by Kets and others (2007) as they comment, "Although few would admit it, many business leaders are like the mythical narcissus they see the person they love most in the world roughly 70% of executives believe they are in the top 25% of their profession in terms of performance."

With this in mind, the need for coaching is clear as is the need to approach this task in a manner most effective for the client. Therefore, in addition to implementation of 360-degree feedback, effective coaching should include time for reflection—allocating time for the leader to freely assimilate information about his or her leadership challenges without the stresses of day-to-day operations. Effective coaching should also use group coaching, whereby all members of the session share their perspectives about themselves and their respective coaching opportunities.

My final point on the subject of coaching is follow-up—an attempt to have group participants follow up with one another on the progress they have or have not made relative to feedback they received (Kets et al. 2007). Often participants don't follow through on something they learned in a coaching classroom when they return to the workplace, particularly when it involves self-reflection and improvement. While some responsibility for lack of follow-through may be attributed to a person's resistance to self-change, according to Goldsmith (2009), as cited by Baban and Ratiu (2012), "Some studies suggest that not all individuals are coachable [and that] coachable individuals are committed to change, [and] have strong motivation to improve their competencies."

Another contributing factor to the lack of follow-through is the leader's return to day-to-day operations and the issues that doing so poses. It is not at all unusual for an executive to return to the proverbial office with the intent of executing certain plans only to find that the picture has changed significantly when he or she arrives. Sales forecasts just went south, a quality issue has occurred in a major manufacturing facility, or a rumor of divestiture has spawned a precipitous company stock sell-off. Although seemingly a bit extreme, these issues come with the territory of executive leadership and cannot be put on the back burner while less threatening concerns are addressed, such as reflecting on received 360-degree feedback.

Despite the commoditized nature of executive coaching and organizational openness to engaging in such developmental processes, there are those who view the need for coaching as a sign of failure or weakness, which may be driven by their introspective views of self-competencies. While superficial coaching may provide a path for leadership-style changes, more visceral behavior modifications require additional insights. And given the personal nature of coaching, it is imperative that coaches be adept in discerning the source of improvement opportunities presented by those being coached. Berglas (2002), as cited by Ellam-Dyson and Palmer (2011), also noted this perspective by "emphasizing how important it is that coaches have the ability to be able to recognize when clients may have deep seated psychological difficulties."

Viewed in a "clinical paradigm" context, an individual's "inner theater" plays a crucial role, not only in how the individual is coached but also in how he or she interprets and responds to such coaching. The "transferential patterns" (actions linked to our past lives) can be powerful and controlling as we are, in essence, forced to relive our past behaviors perpetually (Kets et al. 2007).

As the next section shows, dealing with one's past demons and ghosts often requires much more than external influences. Indeed, a comprehensive understanding of one's operating environment (and the role of emotions in that environment) is a prerequisite for effective leadership.

2.3: Emotional Intelligence and Leadership

It is no secret that intelligence is a fundamental requirement for executive leadership. However, while the technical aspect of intelligence no doubt facilitates critical decision making, personal or emotional intelligence (EI) enables leaders to make the best critical decisions.

Ability to motivate followers to contribute their best in every situation and in all cases is a fundamental property of successful leadership. Often, in order to accomplish this goal, a profound emotional connection between leaders and followers is required. According to Boyatzis and others (2002), "The emotional task of the leader is primal ... it is both the original and most important act of leadership ... [thus] the leader acts as the group's emotional guide." This notion that the leader serves as an emotional guide is key given that, also according to Boyatzis and others (2002), "we rely on connections with other people for our own emotional stability."

It would logically follow then that leaders who possess the capacity to connect at this level are best positioned for success. The importance of employee emotional satisfaction cannot be overstated as it links directly to job performance. In fact, Boyatzis and others (2002) suggest that "employees who feel upbeat will likely go the extra mile to please customers and therefore improve the bottom line."

Given the significant role that EI plays in a leader's overall organizational effectiveness, it is appropriate that we seek to define EI through the lens of various writers. Salovey and Mayer (1990), as cited by Brackett, Rivers, and Salovey (2011), described EI as "the ability to monitor one's own and others' feelings and emotions, to discriminate among them and to use this information to guide one's thinking and actions." This definition suggests the capacity to be in touch not only with the vicissitudes of your thoughts and impressions but also to control how you react to them. Goleman (2000), as cited by Hosein and Yousefi (2012), stated, "The emotional intelligence is an inherent ability and the genes have [an] important role in its creation, but emotional intelligence can grow by training and it needs many efforts and practices." This would suggest that while EI may be attributed to lineage, it is not bound by innate qualities and can therefore be acquired via learned behavior-based methodologies (such as in seminars).

Kets and others (2007) say that "emotional intelligence focuses fundamentally on one's capacity to manage in a social and emotional climate." Within this definition, we are once again reminded of the importance of recognizing the interdependencies of individuals and perhaps more importantly, the leader's awareness of such need for connectedness. Northouse (2013) offers the following comments regarding EI: "As the two words suggest, emotional intelligence has to do with our emotions (affective domain) and thinking (cognitive domain), and the interplay between the two." This definition distinguishes between the two words emotional and intelligence, suggesting that effective use or implementation of EI be predicated on an understanding of emotions resulting from user intellect.

While various authors may offer slightly differing domain descriptors, there are essentially "four domains of EI, self-awareness, self-management, social awareness, and relationship management" (Boyatzis et al. 2002). Self-awareness, as the name implies, suggests that leaders first be cognizant of their emotions and feelings.

Demonstration of proficiency in the area of EI is given by the following example. Assume that an employee named Jack was disappointed with his end-of-year performance review. Such a situation might precipitate a fight, flee, or freeze reaction. Certainly, one response, as damaging as it might be in this situation, would be for him to fight – that is to respond with anger and dissention. While this may seem to be a natural response in this situation, it is not a response consistent with the notion of self-awareness.

In the context of self-awareness, an alternative action might be to first recognize that the different perspectives regarding performance may have resulted from a lack of calibration between subordinate and superior. With this in mind, the conversation may be shifted to how to circumvent a similar situation in the future. The key point here is that the first step for an emotionally intelligent individual is to recognize personal doldrums and to proactively respond (self-manage) in such a manner as to eliminate any further erosion (in the negative sense) of the situation at hand. It is only after Jack is able to be in tune with his emotions (self-awareness) that he will be able to self-manage and subsequently resonate with others.

This perspective is shared by Boyatzis and others (2002) in the following statement: "Self-awareness also plays a crucial role in empathy or sensing how someone else sees a situation." Expanding a bit on empathy, it is important to understand what it is not. It is not about trying to modify your actions such that the masses are sure to like you or about falsely taking on another's feelings as your own. Instead it is about appropriately processing the feelings of others. Boyatzis (2002) and others address the topic of empathy as follows: "Empathy means taking employees' feelings into thoughtful consideration and then making intelligent decisions that work those feelings into the response."

Social awareness, as we have discussed in relation to collectivism, takes into account the emotions of those around us. The leader, in this case having first developed competencies in self-awareness as well as empathy and self-management, is now positioned to perceive and appropriately acknowledge the feelings of others.

Having an awareness of and capacity to manage one's own feelings as well as being able to empathize and connect with others' feelings positions the leader to implement effective relationship management—our last of the four EI domains. More specifically, relationship management is about "authenticity" and how its use may serve to strengthen a leader's connectedness, not only with employees but also with those with whom the leader interacts on a 360-degree basis.

Thus far, we have discussed the four domains of EI and how each might be effectively implemented. However, simply mastering the domains of EI without fully addressing their integration within the leadership realm is incomplete. So leaders should continue to build on existing EI skills and seek to expand the strengths associated with these skills, such as organizational awareness and collaboration. "Having a larger repertoire of emotional intelligence strengths can make a leader more effective because it means that leader is flexible enough to handle the wide-ranging demands of running an organization" (Boyatzis et al. 2002).

There still remains an open question in the area of EI, and that is as follows: How does one develop EI competency? According to Boyatzis and others (2002), "To begin or sustain real development in emotional intelligence, you must first engage that power of your ideal self." This, of course, means to contemplate the person you want to be, which should comprise the elements that invoke the most passion. This profound change requires crafting of a vision reflecting 360-degree interactions and feedback. It is not simply what you, the leader, will be doing but also how you interact with those with whom you make daily contact.

Unfortunately, accurate feedback is often elusive. No one likes to be the bearer of bad news. Subordinates prefer to convey messages that make the boss feel good. Peers sometimes refrain from candor in pursuit of their own agendas; and bosses—believe it or not—often avoid messages that precipitate conflict. Boyatzis and others (2002) offered the following comments in this area: "Rare are those who dare to tell the commanding leader he is too harsh, or to let a leader know he could be more visionary, or more democratic."

With this in mind, it is only through the leader's use of EI skills, namely empathy and awareness, that he or she is able to discover the brutal feedback regarding his or her behavior and how it affects others. The reference to "brutal feedback" may appear a bit harsh and inconsiderate. In fact, providing or receiving such feedback may not be in the best interest of either party if the goal is to appear friendly and unwaveringly collaborative. Viewing it as a sort of hard tactic, Knippenberg and Steensma (2003) stated that "tactics that may be assumed to place a strain on the relationship between agent and target are less frequently employed." This is an important point from a leadership perspective as, although leaders are tasked with motivating workers, which is often viewed synonymously with making everyone feel happy, they should not refrain from providing brutally honest feedback. Important though is that doing so be accomplished with an eye toward empathy as previously discussed. Also important here is awareness and openness to feeling, listening, and thinking and appropriately acting on the inputs received.

Thus far, our discussion regarding EI has focused on the individual level. However, in order to transform the organization, a leader must go beyond self-transformation. He or she is also responsible for the transformation of the team. Key attributes of EI, such as self-awareness, are also applicable at the team level.

Effective use of EI at the team level begins with each member of the team acknowledging the feelings and emotions of every other member. Actions in this area "might also mean creating norms such as listening to everyone's perspective—including that of a lone dissenter—before a decision is made" (Boyatzis et al. 2002). Leaders must master the art of listening to experience organization connectedness and resonance as will be discussed in the next section.

2.4: Resonant Leadership

A not-so-subtle relationship exists between resonant leaders and emotionally intelligent leaders. To an extent, resonant leadership is all about connecting or being in tune with those with whom the leader interacts (e.g., subordinates, peers, and other constituents). Said differently, "when leaders drive emotions positively they bring out everyone's best we call this effect resonance [and EI is] how leaders handle themselves and their relationship" (Boyatzis et al. 2002).

In the lexical sense, resonance is defined as "a reinforcement of sound in a vibrating body caused by waves from another body vibrating at nearly the same rate" (*Merriam-Webster* 2007). From the foregoing definition, the relationship between resonance and EI should be lucid relative to leadership. By definition, motivation is "the act or process of motivating ... a motivating force, stimulus, or influence" (*Merriam-Webster* 2007).

Leadership would be so much easier if all employees showed up motivated to accomplish any assigned task. Unfortunately, the job of employee motivation most often rests with the leader and must be externally sourced. Sure, some help is available to the leader in the form of intrinsic motivation. However, according to Ivancevich and Matteson (1993), intrinsic rewards typically align with one or more of the following categories: "completion—the ability to start and finish [something], achievement—derived when a person reaches a challenging goal, autonomy—right and privilege to make decisions, [and] personal growth—expansion of capabilities." However, what happens in instances where, resulting from job design (for example), the employee is not allowed to complete an assignment or goal before being reallocated to another task or when decisions are handed down rather than allowed or when job stagnation exists? Under these circumstances, the challenge of motivation and thus resonance falls upon the shoulders of the leader.

As I discussed previously, feedback is one tool available to leaders who have served as enablers for boss-subordinate calibration regarding work performance. Indeed, according to DeNisi and Kluger (2000) as well as Gregory, Levy, and Jeffers (2008), as cited by DeBerg, Jarzebowski and Palermo (2012), "Feedback, which is information regarding individuals' current levels of performance, has been shown to influence motivation, job satisfaction and performance." Implemented correctly, feedback, particularly if collected on a 360-degree basis, can offer tremendous returns. According to Wimer and Nowack (1998), as cited by Crispo and Sysinger (2012, "When 360 degree feedback is used appropriately, it can be a very effective tool that can lead to behavioral changes and effectiveness of an individual, group, and organization."

While it is not my intent in this section to discuss the 360-degree feedback instrument in detail (as it was discussed in section 2.2), its usage may certainly facilitate leadership resonance. Whether resonance is enabled through motivation or otherwise, tuning in to the resonant frequency of multiple followers, while certainly doable, is not accomplished without tremendous effort and persistence, which can be extremely exhausting, and if left unaddressed, leader burnout is inevitable.

How then should a leader continuously replenish the well—the source of motivation, guidance and emotional drain? Boyatzis and McKee (2005) believe that this is accomplished via a "cycle of sacrifice and renewal that must be regulated to maintain resonance." The type of stress precipitating the need for renewal is termed "power stress" and is the source for dissonance. Contributions to this stress type are provided by ambiguity and requirements for complex decision making. "Firefighting" is another source of this type of stress, and in some situations, leaders may become physically ill as a result of the day-to-day battles.

The principal issue with power stress is not necessarily the effect experienced while in the heat of the battle. It is instead "too little recovery time," which results from leaders "failing to manage the cycle of sacrifice and renewal" (Boyatzis and McKee 2005). It is this process of renewal that allows leaders to sustain connectedness within and among the organization.

Leaders are continually being assessed, analyzed, and scrutinized. Not only are company owners (shareholders) seeking optimal returns, boards of directors are also demanding unprecedented results while employees are looking to be coached, promoted, complimented, and supported, not to mention given clemency regarding mistakes.

It is not unrealistic to assume that leaders contemplate the antagonizing aspects of these events ahead of their occurrence. This perspective is supported by Martin (1997), as cited by Boyatzis and McKee (2005), in the following comment: "Humans have what many consider a unique ability to create their own stress by merely anticipating stress-inducing situations." In the most fundamental sense, the cycle of sacrifice and renewal has been presented to each of us from day one. As infants, we might be encouraged to accomplish a goal or task only later to be rewarded with something worthy of the sacrifice—a sort of renewal for our efforts.

Another similar example is the typical sports drink commercial depicting an athlete accomplishing a feat through physical exertion only to later be rewarded with a bottle of colored liquid consumed while in a position symbolic of achievement and gratification. In leadership, mental stress, unlike physical stress as described previously, may be directly related to psychological health and well-being. Under the conditions of power stress, the "sympathetic nervous system (SNS)" is aroused, which precipitates the fight-or-flight response as discussed in the section regarding EI.

Combinations of certain types of stress encountered in the day-to-day leadership circle are "said to increase the allostatic load," which can result in severe health issues. Under these stressful conditions, increases in "multiple neurotransmitters" occur, which may also result in increased blood pressure (Boyatzis and McKee 2005). While power stress implications may be most profoundly realized in the SNS, the "parasympathetic nervous system (PSNS)," when appropriately stimulated, is the system responsible for recovery from any such stressful condition. Such renewing stimulants may include "hope," "compassion," and "meditation," acting as a sort of "antidote to stress" (Boyatzis and McKee 2005).

Leadership is not a job for the meek at heart. Not only are sacrifice and renewal integral to long-term effectiveness, but the leadership responsibility also requires self-discipline, a willingness to make the tough calls, an almost uncanny knack for providing brutally honest feedback, and perhaps most importantly, the ability to feel comfortable feeling uncomfortable. However, despite the vastness of a leader's soft and hard skill repertoire, nothing precipitates more respect from a leader than his or her credibility as I discuss in the next section.

2.5: Transformational Leadership and Credibility

Despite the best business school preparation, only experience in the field can prepare an executive for the vicissitudes of leadership. Transcending these ups and downs of leadership is earned credibility, which often serves as a prerequisite for leadership effectiveness. While the principal goal of this section is to discuss leadership credibility through transformational applications, primarily to allow for reader comprehension, I also briefly address the transactional leadership (XL) and transformational leadership (TL) styles.

Nystedt (1997), as cited by Korner and Nordvik (2004), suggests that "behavioral styles have been elaborated into constructs such as charismatic, transactional, transformational and visionary leadership." Focusing on TL and XL styles, we find, according to Cilliers and others

(2008), the following distinguishing characteristics: "[TL]—idealized influence, implies that followers respect, admire, and trust the leader and emulate his or her behavior, assume his or her values, and are committed to achieving his or her vision and making sacrifices in this regard ... [XL]—involves a social exchange process where the leader clarifies what the followers need to do as their part of a transaction (successfully complete the task) to receive a reward or avoidance of punishment (satisfaction of the followers' needs) that is contingent on the fulfillment of the transaction (satisfying the leader's needs)." It might be argued that the characterization of XL is predicated on certain aspects of Maslow's needs hierarchy as I will discuss in a subsequent section.

Referring once more to TL, which is built on openness and engagement, Lo, Min, and Ramayah (2009) wrote, "Transformational leaders [have] a more significant relationship with organizational commitment." Through motivation and workforce engagement, TL builds equity in the form of employee loyalty, which serves the entire organization and its constituents.

One very simple yet often illusive TL practice that facilitates organizational engagement is listening to the employees. In turn, this enables four dimensions of effectiveness. First, the leader is able to gain an understanding of how employees view the world around them and thus how they might interpret direction provided to them. Second, the leader is able to begin the process of connectedness (previously discussed), which enables the engagement process. Third, the leader gains the respect of employees because they now feel that someone—one quite powerful in the eyes of the organization—cares about what they have to say. Finally, the leader gains insight as to what is really happening within the organization, and depending on the employee's organizational hierarchy, critical operational details that might otherwise be overlooked are now made available to the leader. Listening to and engaging employees also sets the groundwork for the leader to execute the "five practices of exemplary leadership: model the way, inspire a shared vision, challenge the process, enable others to act, and encourage the heart" (Kouzes and Posner 2007).

To this point, we have discussed benefits resulting from leader engagement with the organization. However, engagement alone is not the panacea for leadership effectiveness. Such interactions, particularly when considering TL, are assumed to be authentic. With this in mind, resulting from the leader's behavior, an increased level of organizational integrity and morality should be realized, and thus, leadership credibility will increase too. Indeed, according to leadership attribute survey results referenced by Kouzes and Posner (2007), "For people to follow someone, the majority of constituents believe the leader must be honest." This perspective is appropriately aligned with characteristics of transformational leaders. Indeed, according to Burns (1978), as cited by Plinio (2010), "In transforming leadership, persons engage with others in such a way that leaders and followers raise one another to higher levels of motivation and morality."

Integrity and honesty are the building blocks for leadership credibility which, according to Kouzes and Posner (2007), requires leaders to "practice what they preach, walk the talk, actions are consistent with their words, put their money where their mouth is, follow through on promises, and do what they say they will do."

Another key attribute of TL, as mentioned previously, has to do with creating a shared vision. As viewed here, a shared vision is one whereby the organization doesn't simply march to the drumbeat but also picks up and carries the torch in one accord with ownership as though the vision was crafted from the bottom up. Kouzes and Posner (2007) suggest that "visions are ideals" and that as such, "they're expressions of optimism" that should "appeal to common ideals."

If TL attributes are correctly imparted, the organization should assume the leader's values. The focus here is on shared and synchronized values, which "are the foundations for building productive and genuine working relationships." As a result of this approach, "tremendous energy is generated when individual, group, and organizational values are in synch" (Kouzes and Posner 2007).

Still another attribute of the transformational leader is trust. This important leadership attribute serves as a critical factor for leadership efficiency and resource optimization. Viewed in this way, when a leader assigns work within an organization, that leader can do so with utter reliance on the worker to accomplish the task or conversely, with follow-up and questioning in such a way as to micromanage the worker. In the latter case, work efficiency is reduced in two areas. First, the leader is now allocating time that could otherwise be used to accomplish other more strategic activities, and second, the worker is now focused on the next intervening moment initiated by the leader and reverts to a sort of wait-for-direction mode, effectively slowing down the processing of the received input. (Refer to figure 1-1, a description of leadership systematization.)

Trust in either direction (trusting others or being trusted by others) is an essential component of effective leadership. Kouzes and Posner (2007) share this perspective with the following comments: "At the heart of collaboration is trust ... without trust you cannot lead ... you cannot get extraordinary things done."

Trust and engagement add to the list of leader credentials and aid a leader in moving toward the state of credibility. However, new leaders are often expected to manage more than the status quo. They are expected to convert lost revenues to new profits, to replace inefficiencies with productive operations, to modify and eliminate in some cases existing outdated infrastructure. In effect, leaders are expected to initiate and bring about profound and sustainable change. The transformational leader is adept at delivering in this regard. This perspective is shared by Crant and Bateman (2000), as cited by Belschak, Deanne and Hartog (2012) in the following comment: "Transformational leaders are more change oriented and proactive themselves and thus may act as role models."

In the quest for credibility, perhaps the most assumed quality that a leader possesses is the intellectual wherewithal to stimulate the thoughts and creativity of others. Indeed, positioned correctly, learning is fun, and employees do well to know that they can be taught new strategies, approaches, and ways of thinking. Transformational leaders thrive on intellectual stimulation as supported by the following comment offered by Bass (1985), Avolio and Bass (1988, 1990a, 1990b), and Howell and Avolio (1993), as cited by Atwater, Avolio (1996), and Bass:

"Transformational leadership has been shown to include inspirational [and] intellectual stimulation."

Finally, leadership credibility is also about caring for and supporting those whom the leader is entrusted to lead. You should not assume that listening (discussed earlier) is necessarily synonymous with caring as listening alone could in some instances represent a purely perfunctory event aimed solely at advancing the leader's agenda. Transformational leaders gain credibility through sincere actions and caring. According to Bass (1985, 1998), as cited by Liu, Siu, and Shi (2010), "Transformational leaders ... show their concern for their employees' individual needs for growth and development." Leadership style, whether it be TL, XL, or otherwise, is not achieved without cognition.

In the next section, we discuss the psychological implications of leadership.

2.6: The Role of Psychology in Leadership

Let's begin by recalling the definition of psychology, which may be summed up as the characterization of human behavior. Perhaps leadership can be viewed as an attempt to positively influence follower cognition and emotion such that the followers feel good about themselves and what they can accomplish and are thus motivated to execute their jobs with quality.

We have discussed several approaches to leadership. We have visited theoretical leadership models and have discussed available leadership tools. What remains an open area for discussion is how the cognitive process functions while interpreting the various leadership approaches. Why is it important to understand the role of the cognitive process in leadership? Or said differently, why is leadership motivation necessary at all? One response is that "it has been estimated that organizations suffer up to $370 billion in lost productivity every year in the United States alone due to workers not feeling engaged" (Lawrence 2011). Thus, an understanding of the cognitive process coupled with the appropriate leadership motivation offers the potential for tremendous returns.

While intellect is a prerequisite for good leadership, intellect "alone will not make a leader; leaders execute a vision by motivating, guiding, inspiring, listening, persuading—and, most crucially, through creating resonance" (Boyatzis et al. 2002). The section of the brain that controls and provides intellect is separate from the section that guides emotion. However, under the appropriate circumstances, the two are integrated such that emotion takes over and in effect rules (Boyatzis et al. 2002). It should not be surprising that the brain succumbs to emotion because emotion serves as the on-off switch for responding to stressful situations—such as to the performance review I discussed in section 2.3 of this chapter. Indeed, the "thinking brain evolved from the limbic brain and continues to take orders from it when we perceive a threat or are under stress" (Boyatzis et al. 2002).

The problem with the brain is that it originally developed to protect us from physical environmental threats. The brain is not innately structured to handle the stresses associated with a bad performance review or to respond to office politics. Importantly, though, one's ability to circumvent sudden and perhaps unwanted reactions in such situations is attributed to the brain's executive center or neurologically speaking, the prefrontal area of the brain. The circuitry responsible for actions executed by the executive center also controls drives and impulses. Unlike the process required for technical learning and skill development, EI-oriented skills are "best learned through motivation, extended practice and feedback."

The emotional or limbic side of the brain is far less developed than the thinking brain (the side that aids in technical learning). As a consequence, a great deal of limbic learning results from repeated exposure and early behavior introductions. This early information is accessed through maturity and in a manner, "as if it were factual." Thus, decision making is predicated on our cultural preferences and biases, which could very easily result in disconnects with contemporary environments.

As leaders, the same bias carries forward in the direction that we provide to others, and as followers, our cultural biases serve as the basis for how we interpret information received from leaders. These comments, however, are not at all intended to suggest that the brain cannot be taught to act in an emotionally responsible manner because "human brains can create new neural tissue as well as pathways throughout adulthood" (Boyatzis et al. 2002; Bailey 2007).

When a new idea is presented, say a new sales goal or mission, the prefrontal area of the brain is asked to consider this new information and compare and contrast it to prior similar information, such as the old sales goals or mission. "The basal ganglia" part of the brain is engaged for routine activities, such as driving a car, and represents the part of the brain that stores habits and routines. Under change conditions, however, such as driving a car on the left side of the road, the prefrontal cortex becomes active.

This same cognitive dynamic occurs when employees are exposed to organizational stresses and change. The norm is for our brains to gravitate toward things with which we are both familiar and comfortable. Thus, under change conditions, especially conditions that deviate from expectations, the brain emits strong signals that reflect acknowledgment of the deviation. The part of the brain in which these deviation signal emissions occur is the orbital frontal cortex as part of the brain's fear circuitry. The occurrence of such signal emissions can precipitate emotional or impulsive responses, propelling us to fight-or-flight behaviors.

Some have posited that changed behavior can be accomplished via behaviorism-oriented approaches. One example would be to associate a desired behavior with a reward as in the so-called carrot-and-stick approach, which, despite its convincing appeal, has been proven by clinical research to be ineffective (Rock and Schwartz 2007; Bailey 2007).

Given the complexities of the human brain, it is imperative that effective leaders make appropriate connections with those whom they lead toward change. (Recall discussions regarding resonance.) Often communication serves as the enabler for such connections. It must therefore be executed with the utmost care and scrutiny.

This perspective is shared by Kussrow (2001), who writes, "Since it is people's brains that leaders try to influence ... it follows that it is critical that the individual being [led] accurately interprets what the leader intended to communicate." Within this communication should be options and choices for followers—a sort of participative versus dictatorial style of leadership. The reason for this is that "humans have a social brain that loves to anticipate, to be given choices" (Kussrow 2001).

Despite the brain's desire for expectation and variation, humans bring to the table old habits that are often difficult to change, including pessimistic views about planning and executing their daily responsibilities, about meeting project deadlines, about solving old problems, and about proactively identifying and resolving latent problems. These attitudes are the norm. Indeed, "changing behavior is hard, even for individuals and even when new habits can mean the difference between life and death" (Rock and Schwartz 2007).

Every organization comprises individuals with disparate habits and varying levels of organizational commitment. Thus, it is not at all surprising that any attempt to change an organization's mind-set may be extremely difficult. "Organizational transformation that takes into account the physiological nature of the brain, and the ways in which it predisposes people to resist some forms of leadership and accept others [may offer the best chance for success]" (Rock and Schwartz 2007).

In my previous discussion regarding an annual performance review, the focus was on the disappointed recipient (the subordinate). It suggested that the opportunity for the application of EI rested with the follower. However, given the foregoing discussion regarding the limbic system and its relationship to EI, an alternative perspective would be to view the responsibility for a successful discussion to rest with the deliverer (the superior). Said differently, the superior's awareness of how the limbic system functions coupled with his or her maturity in EI affords this individual the opportunity to change the conversation at the outset so that the subordinate needn't encounter the fight, flight, or freeze syndrome.

Leaders should also be cognizant of the four basic drives of leadership. According to Lawrence (2011), these include the drive to acquire, defend, bond, and comprehend. While the drives to acquire and defend are principally concerned with survival and self-preservation, the drives to bond and comprehend focus more on relationship building and individual perception respectively.

If something does not progress according to plans or expectations, such as sales results not meeting forecast levels, the "drive to defend" may move a leader to overlook key information that explains why sales fell short. Recognition of your position as a leader in such a situation will allow you to engage your executive center to respond appropriately.

Building on this point, according to Lawrence (2011), "to be effective, leaders must take into account how the four drives affect the following group characteristics: purpose, competencies, trust building [and] motivation." Returning to the performance review discussion, an employee could perceive the absence of a good review as a threat to the right to acquire. This could give rise to the development of barriers to building trust with the leader.

Another perspective on human requirements is seen in Maslow's hierarchy of needs (mentioned previously). According to Ivancevich and Matteson (1993), Maslow's five-stage model includes the following five human needs (listed hierarchically): 1) physiological; 2) safety and security; 3) belongingness, social, and love; 4) esteem; and 5) self-actualization. Returning to the performance review discussion, the threat to an employee's right to acquire (in this case, a good performance rating) could also represent a threat to physical needs, including food and shelter.

Food and shelter, of course, represent components that are essential for meeting human physiological needs. You might question the relevance of physical or biological needs to psychology. The bearing of such is actually quite simple and can be explained by recalling a natural disaster—Hurricane Katrina. Coverage of the hurricane aftermath showed the victims as missing the most fundamental of human needs, including "food, drink, shelter and relief from pain" (Ivancevich and Matteson 1993). Yet the degree to which these needs no longer exist is somewhat psychological and relative. For example, some of the victims, despite having lost homes, were provided shelter and food by philanthropic organizations such as the Red Cross. So the reality is that while the victims no doubt suffered hardship, in the purest sense of Maslow's hierarchy of needs, the first level in the needs hierarchy continued to be met, albeit with the assistance of others.

This disconnected observation (failure to recognize fulfillment of Maslow's first-order need) may be attributed to perception based on the victim's frame of reference before the hurricane. Psychological factors may certainly give rise to feelings of pessimism. Indeed, many third world cultures would find great solace in the levels of provisions after the hurricane (also addressing Maslow's first hierarchical need of food, water, and shelter) as were afforded to persons impacted by Hurricane Katrina. Psychology affects perception, and perception in turn is linked to motivation. Therefore, effective leadership must also address the notion of perceptions and instill feelings of optimism.

I certainly do not aim to downplay the disruption caused by Hurricane Katrina. Without a doubt, enduring it was a tremendous psychological and biological injustice to everyone who was impacted. Rather my objective was to illustrate the power of perception and how important it is for leaders to be cognizant of psychological influences and motivations. Having established the link between perceptions, psychology, and effective leadership, the next section discusses construct-based methods for the identification of certain leader attributes that may also be predictors of effective leadership.

2.7: Predicting Leadership Behavior

Although advancement has been made in predicting leadership behavior based on psychometric modeling, the fundamental concept is not new. According to Lynam and Miller (2001), "Since its inception, the field of personality research has been concerned with identifying the basic traits that serve as the building blocks of personality." Among some of the most researched behavioral models are Five-Factor Model (FFM) (McCrae and Costa 1990), Three-Factor Model (PEN) (Eysenck 1977), Three-Factor Model (Tellegen 1985), and Temperament and Character Model (Cloninger et al. 1993; Lynam and Miller 2001).

Lynam and Miller (2001), also suggest that the basis for these models ranges from "lexical hypothesis" associated with the FFM to "factor analysis and mood scales" used by Tellegen to "biological/pharmacological" associated with the Cloninger and Eysenck models There is also the Myers-Briggs Type Indicator (MBTI), which was designed by Briggs and Myers and, according to Carlson (1985), "is a test designed to implement ... theory type ... therefore, like the projective techniques, the MBTI is closely allied with psychodynamic thought, at least in its original conception."

Expanding discussions regarding the FFM, we find, according to Costa and McCrae (1992), as cited by Kornor and Nordvik (2004), it is "a hierarchical model of personality traits with five big traits called domains on the top, that is, Neuroticism, Extraversion, Openness, Agreeableness, [and] Conscientiousness." According to Levine and Raynor (2006), each of these five domains is further defined as follows: "Openness—refers to intelligent, imaginative, curious, flexible and broad minded; conscientiousness—refers to striving for competence and achievement, and being self-disciplined, orderly, reliable, and deliberative; extraversion—refers to enjoying the company of others, and being active, talkative, assertive and seeking stimulation; agreeableness—refers to being courteous, good natured, cooperative, tolerant, and compassionate rather than antagonistic; neuroticism—refers to easily experiencing unpleasant and negative emotions, such as fear, anxiousness, pessimism, sadness, and insecurity."

There has been much discussion regarding the FFM and its ability to predict leadership or other behavior-based traits (such as conscientiousness). To this end, according to Srivastava (2010), "my thesis is that we will never really understand the Five-Factor Model until we more fully come to grips with the scientific implication of lexical hypothesis." However, as also pointed out by Srivastava (2010), "The Five-Factor- Model is first and foremost a model of social perceptions." Srivastava's position is somewhat supported by Saucier and Goldberg (1996) as they stated, "The big five [FFM] are dimensions of perceived personality." Moreover, D. W. Fiske wrote, as also cited by Srivastava

(2010), that the FFM is useful for "the analysis of how people perceive people and what words they use in formulating such perceptions."

Considering the breadth of the English language, it is perhaps unthinkable to consider that the lexical approach may be constrained in its capacity to fully describe personality traits. Yet words are just that, and how they are interpreted from one human being to the other is not as consistent or black-and-white as their use might suggest.

An analogy to this thinking is offered by Palmer (1999) and Adelson (1990), as cited by Srivastava (2010), in the following statement: "But color perceptions have unique qualities and special relationships that do not purely reflect the extra human physical world, and the perceptual processes that ordinarily help us perceive color can lead to errors under some conditions." Inconsistencies of interpretation notwithstanding, the comments offered by Srivastava, as well as his cited sources in this area, precipitate recollection of an adage that we have all heard, namely that "perception is reality." Thus, valid as arguments may be on both sides, the FFM construct is quite relevant to the core of leadership style-measuring instruments as it evidences the capacity to offer individual behavior validity through observation.

Having laid a comprehensive foundation for leadership, leadership styles, and measurement constructs, the subsequent chapters in this book provide opportunities for integration and application of the preceding review. They also seek to broaden the leadership repertoire via practical corporate simulated scenarios.

CHAPTER 3

Power and Leadership

ACCORDING TO NORTHOUSE (2013), "The concept of power is related to leadership because it is part of the influence process ... power is the potential to influence." I'd like to start this section in a somewhat unconventional manner. What I mean by this is that I'd like for you to take just a moment to grab a sheet of paper or something else that can be used to cover up everything on this page following this first paragraph—that is, the one that you are currently reading. No, do not look ahead to the next paragraphs. If you have purchased this book via e-book, you may want to scroll down so that you only see this paragraph. Okay, ready? At this point, what I'd like for you to do is to consider the word *power*. What comes to mind when you think of power? I know some of you may be thinking about the word *powerful*, while others may be thinking of people and powerful positions such as the president of the United States or the CEO of your corporation. Some of you may simply be thinking of your boss or someone you revere such as a coach or teacher or historical figure. Take a few moments to consider this question, and feel free to jot down your thoughts. After you have completed this brief exercise, you may uncover this page and read the rest of this chapter.

I must say that with an electrical engineering background, when I hear the word power, I immediately begin to think about something referred to as Ohm's law, where power is equal to voltage in a circuit multiplied by the current in that same circuit ($P = V \times I$). But of course, that's simply the view of someone grounded in (no pun intended) and fascinated by the physics of electrical circuitry. According to *Merriam-Webster (2007)*, the word *powerful* is defined as "having great power, prestige or influence." It should be clear then that one cannot be perceived as powerful in the absence of power. Thus, we must return to our original question. *What is power? Merriam-Webster (2007)* defines power as "the ability to act or produce an effect ... a position of ascendance over others ... one that has control or authority." Social psychologists French and Raven (1959) viewed power in a slightly different context, relating it to society. In sum, they suggested that power has a great deal to do with social influence. They then defined social influence as bringing about "a change in a person's behavior, opinions, attitudes goals, needs, values and all other aspects of the person's psychological field" caused by another person or group. French and Raven further defined power as "the ability to influence others."

Early in my career, having extensive expertise in electronic circuit design brought attention to my capability as an electrical engineer. But what does expertise have to do with power? Well, everything. You see, power can also be viewed in terms of power bases. What I mean by this is that there are different types of power. French and Raven (1959) expanded their views to suggest that there were actually five types of power, which they refer to as expert power, legitimate power, coercive power, reward power, and referent power. At the time, I was perceived to have expertise in electronic circuit design. Consequently, that was my expert power. Each of you, I am sure, has a similar expertise—something that you do just a little bit better than anyone else on your team. At the same time, each of your team members has expertise in a certain area that he or she does just a little bit better than you. However, at this moment, it's all about you and what you bring to the table. The big question now is what exactly is your expert power? So I'd like you to take a moment and respond to the following statements: List three things that you do well. List two things that you do very well. List one thing that you do better than anyone else in your work group. This is something that you are perceived to be an expert at doing.

If you got through the first two bullets okay but could not answer the third one, then either you are too modest, or you may need to beef up your skill set. It could also be that you are functioning in a role or job that is not fully compatible with what you bring to the table. There's no need to panic, however; just about everyone could benefit from further developing their existing skill sets. So let's go through an example. Suppose that you are an entrepreneur. One of the things that you might do well is provide good response to customer requests. You might be very good at working with your customers to determine the best product or service that you offer for their specific needs. What you might do better than any of your competitors is assuring that the purchased product meets their needs by providing on-site start-up and support at no extra charge. Perhaps you have product start-up and calibration expertise, or you understand the customer's process better than your competitors. Thus, while the competition simply sells products, you guarantee that the products you sell will operate as desired and fully meet customer needs. You could say that this is your competitive leadership advantage, as it also represents your expertise in the area.

As another example, suppose that you're a member of the team charged with trying to determine the cause or causes for consumer complaints. As a member of the team along with other team members, you might be good at analyzing reports. This will be one of the three things that you do well. Taking this a step further, you may also be able to look at data—consumer complaint data in this case—and very quickly identify trends. This is one of the things that you might do very well. Now when it comes to the one thing that you do better than anybody else in your work group, it might be that you are able to enter data into an elaborate spreadsheet and apply high-power statistics to figure out if changes in the process affect the product quality as measured by consumer complaints. This then is your expert power.

The importance of leveraging your power and in particular expert power cannot be overstated. Indeed, according to Allison and others (2015), "Empirical results show that expert leaders are associated with better organizational performance in a number of settings (e.g. universities, hospitals, high-technology industries, sports)." The article continues that "the theory proposes the existence of a first-order requirement—that leaders should have expert knowledge in the core business of the organizations they are to lead." Given the previous discussion on the topic, I would submit that these areas only touch the surface regarding the applicability and value add of this form of power. Although engineers typically do well in exhibiting expert power (particularly in the technical arena), they are not generally viewed as possessing key leadership competencies. Weingardt (1994) articulated this perspective by stating, "The public perception of engineers—and perception is more powerful than reality—is that we do not have leadership qualities." A critical reference used here is "perception," which, as we will see later in this chapter, is in itself quite powerful.

Let's shift now to discuss legitimate power. This is perhaps the most frequently viewed form of power. Position power is a very obvious form of legitimate power. For example, your boss has legitimate power, and as such, he or she can determine your performance rating. A football coach has legitimate power, and he or she can bench a player at his or her discretion. Your teacher or professor has legitimate power and can therefore allocate an A, B, or other grade on a paper that you submit. As you can see, an individual's position may give legitimacy to his or her power. Positions of authority, while perhaps the most common, are not the only sources of legitimate power. Such may also result from one being (or feeling) obligated to provide some service or action to another. For example, if I commit to trimming the shrubs around my house, my wife is then empowered to ensure that I follow through on that commitment. (As a side note, when this occurs, she is not at all hesitant to exercise her legitimate power. If you don't believe me, just ask her.)

Another way of looking at legitimate power is from a cultural perspective. Viewed in this way, we see that family members are also conferred some level of power. As a child, if we were disciplined by our aunt or uncle, our parents would offer similar discipline as well. Said differently, we were liable for spankings from parents, aunts, uncles, and just about any other family members. Thus, the extended family was granted a level of disciplinary power legitimacy simply because of cultural beliefs and norms. Through our election and democratic process, we legitimize our elected officials and therefore grant them power. Okay, now hold on tight for this one. Not everyone in a legitimate position actually has power. How can that be? One quite obvious example is that of a lame-duck president. That is a president whose successor has already been elected. In this scenario, the incumbent enjoys only significantly reduced levels of power although he or she may remain in the position for several weeks or months postelection of the successor. At this point, some of you are no doubt wondering if you have legitimate power. After all, you may be serving in an entry-level position at your company. Actually, you have a great deal of power conferred to you by virtue of your job description. This description grants you all sorts of legitimate power. The question becomes thus: When is the last time that you reviewed your job description, and are you fully leveraging its conferred power to accomplish great things?

The discussion of legitimate power—particularly that conferred through position—is a great lead-in for discussing our next type of power (coercive power). As the name implies, this type of power is all about having the ability to force someone (e.g., a subordinate) to execute an order or direction to avoid punishment for lack of compliance. This punishment could include many things, such as a low performance evaluation, reduced pay, bias in allocating assignments, and many other punitive issues, including but not limited to termination. I think it goes without saying that this form of power, despite its teeth, should be one of the least desired forms of action by the one in charge. Coercive power is not always about doling out punishment inasmuch as a person's existing job or income is threatened. It could also result from a boss or person in charge withholding bonuses or other privileges, perquisites, or what have you that the employee would otherwise receive. Despite the negative connotation associated with the use of coercive power, there are situations where the use of this power is quite warranted. As an example, if someone that you knew, such as your peer, was stealing company information or secrets, you might issue a firm warning to that individual suggesting that unless he or she discontinues the behavior, you would have no choice but to turn him or her in to the proper authorities. This might particularly be of interest to you if, as a result of the employee's stealing of company information, the company was losing market share and therefore threatening to reduce employee pay. Another example in this scenario might be whereby the president of the United States, in an attempt to maintain a safe and secure nation, threatens the use of military action against a rogue nation viewed as a threat to our national security. Thus, use of coercive power in this instance may preserve our way of life.

It is logical at this point to discuss reward power. It could be argued that applications of coercive power are simply the withholding of some form of reward power. Said differently, if you are not being punished, that in itself may be a form of reward. Yet the context for reward power as discussed here simply focuses on that which is given to someone based on the giver's perception that a reward of some sort is due. And returning to the examples previously outlined, a great performance evaluation, a pay raise, or the opportunity to receive a highly visible assignment all fall into the category of reward power. Another way of looking at reward power is to consider the example of being accepted. During my younger years, I had a strong penchant for competition. Consequently, when my pickup basketball or football team won several games, I would swap places with a member of the losing team so as to compete with my former teammates. My acceptance by the losing team represented a form of reward. After all, I was one of the opponents previously serving to defeat them. Thus, as you see with this example, the mere acceptance by others, whether it is on a team or otherwise, does indeed represent a form of reward.

Shifting now to our final power base, namely referent power, you have no doubt heard the adage "Birds of a feather flock together." The point being made here is that individuals who can readily identify with one another are likely to prefer one another's company versus those with whom there is much less in common. How many times have you observed someone you felt would be a great mentor or adviser? In doing so, the mentor or adviser—from your perspective—most likely exhibited referent power. As a teenager, I progressed through various assignments while working at a restaurant, which ultimately led to achieving the assistant manager position. While the progression was certainly driven largely by my passion to achieve greater economic status, it was also driven, in part, by my desire to emulate my boss at the time. To me back then, he epitomized possession of referent power. He played the guitar, drove a GTO, and had an overall great personality. Indeed, I purchased a GTO and played the guitar (or at least attempted to play the guitar), and we both enjoyed some of the same artists, including Manfred Mann's Earth Band, Boston, the Eagles, and many others. To this day, I think if our paths crossed, there would still be many things that we would have in common. During the 2016 presidential campaign, it was widely reported that one candidate stated, "I could stand in the middle of Fifth Avenue and shoot somebody, and I wouldn't lose voters." Notwithstanding its validity, this statement implies a level of referent power—at least as perceived by the candidate. In this particular case, messaging played a key role in the perceived level of referent power. Said differently, some people felt a connection with the candidate as they may have also perceived the candidate to empathize with their key concerns or beliefs.

It may be useful for you to take a moment and conduct a self-attribute inventory—that is, the things about you that others perceive to be appealing or that otherwise might draw others toward you or your circle. With this knowledge, you will be empowered to fully leverage your referent power to influence (in a positive way) those in your circle. Taking self-inventory might also indicate that you may not be fully demonstrating the referent behaviors as you thought, or perhaps in a more likely scenario, you may just not be demonstrating them to the degree that others are likely to fully recognize. Referent power is not limited by individual attributes. Certain groups, organizations, and other multimember associations that wield quite a bit of intrigue also exhibit referent power. As an example, upon completion of my PhD program in engineering management, I was asked to join the Engineering Management Honor Society. Because a prerequisite for receiving an invitation was stellar performance in all academic work attempted, membership in this organization was the envy of many. So now the question becomes thus: What role do you see referent power playing in your life?

There is one other point to be made here. We have discussed power under the premise that those who possess it are also perceived to possess it. This is a very critical assumption and is perhaps best understood through a humorous Christmas tale. As a child, my siblings and I often watched the 1964 TV movie *Rudolph the Red-Nosed Reindeer*. For those of you who may not be familiar with the movie, it was about a reindeer and other so-called misfits who later joined forces and turned out to be the heroes. Among all the villains in the movie was the abominable snow monster. Initially, he was perceived to be quite intimidating and ferocious. You might say that all the other characters viewed him as quite powerful. That was, of course, until the misfit dentist (Hermey) pulled all the abominable snow monster's teeth, and Yukon Cornelius (another character in the movie) reformed the snow monster. Thus, what was once perceived to be a ferocious and antagonistic opponent was now viewed as a helpful friend who, leveraging his height, placed a gold star on top of a very tall Christmas tree. There is still a bit more to the story, but I'll leave to you to review it. The point, however, is that once the townspeople no longer perceived the snow monster as being a threat, his power level was reduced significantly. When considering coercive power, that which may be viewed as punishment for one may be viewed as no punishment at all for someone else. As an example, I previously spoke of the performance evaluation and how someone of position power may allocate a low performance rating as a form of punishment. However, an employee not wishing to grow in organizational status or exert anything beyond that which is absolutely required to get the job done may well be satisfied with a low performance rating. It's all about perception. At the same time, an employee who is accustomed to receiving the highest weighting of a performance evaluation and who then transfers to another job and receives a just above-average performance rating may view the "just above-the-average reward" as little to no award at all. Again, the transferring employee had a different perception of reward that was in this case triggered by his or her past experiences. Similarly, expert power is all about perception. Rylander (2015) addressed this view by stating, "To say that a coach has … expert power is to say that the athletes perceive the coach as someone who can provide the athlete, or team, with the knowledge they need to develop or to perform well, regardless of whether they actually have this knowledge." The key takeaway from this discussion regarding power and perception is that when others perceive that you have power, they, in essence, give you power.

I'd like to summarize this entire chapter with the following comments offered by Frederic Labarre (NATO Defense College in Rome, Italy) in a geopolitical context. In discussion what he viewed as "some of the staples of state power" [he wrote] "power is essential in achieving and maintaining independence." While I certainly recognize that you are not a state in your own right, we should note that states comprise corporations, businesses, hospitals, and academic institutions. You are no doubt affiliated with one or more of those in some way. Thus, it could be argued that you (in aggregation) make up the states. It follows then that power should not be limited to the states but that you should take back and leverage the power of your power!

Thus far, we have discussed the five power bases and how each might be leveraged to afford the individual maximum potential for mobility. Yet, we have only discussed the topic of power in the general sense. Figure 3-1 reiterates this discussion. In this section we will focus specifically on how leaders might leverage key power components. Why is this of significance? According to Falbe and Yukl (1992), "To be effective, managers must be able to use power to motivate subordinates, gain support from peers, and influence superiors to provide resources." Continuing along these same lines of thinking, Thinaut and Kelley (1959) defined power as "the control over valued resources or, equivalently, the capacity to affect outcomes" (Tjosvold 1995). In figure 3-1, we have listed the five bases of power as a sort of interdependent system whereby each individual attribute is linked to and dependent upon another. In leadership, the use of all power attributes, including coercive, is necessary depending on the situation.

Figure 3-1: Bases of Power

Let's digress for a moment. In chapter 2, I introduced and explained the difference between transactional and transformational leadership styles. Abbreviating that discussion here, while transactional leaders are more concerned with command and control or contingent reward, transformational leaders focus on inspirational motivation, idealized influence and behaviors, individual consideration, and intellectual stimulation.

Transformational leadership (TL) has been identified as the preferred leadership style. Indeed, according to Avolio and Bass (2004),

> "When all levels of managers, students, and project leaders around the world were asked to describe the characteristics and behaviors of the most effective leaders with whom they had worked in the past, the characterizations were more transformational than transactional. Among the specific descriptors used for these leaders were "inspirational, intellectually stimulating, challenging, visionary, development oriented, and determined to maximize performance."

Returning now to the five power bases (expert, legitimate, coercive, reward, and referent), we should now be positioned to suggest how each may align with transformational and transactional leadership. We know from our previous discussion that legitimate power is often associated with position and is thus somewhat transactional in nature. As an example, consider a time when you as a boss assign a certain performance evaluation rating. Such is executed based on your legitimate right to do so. It could further be argued that by virtue of assigning the rating, particularly if the rating is above average, the boss is also exhibiting an extended form of reward power. And to the degree that the rating reflects above-average accomplishment of organizational goals and objectives, it, too, reflects contingent reward and is thus transactional. Likewise, the threat of punishment for not achieving organizational goals and objectives could be viewed as exercising a form of coercive power. A graphical representation of these transactional power bases is captured in figure 3-2.

Figure 3-2: Transactional Leadership Power Bases

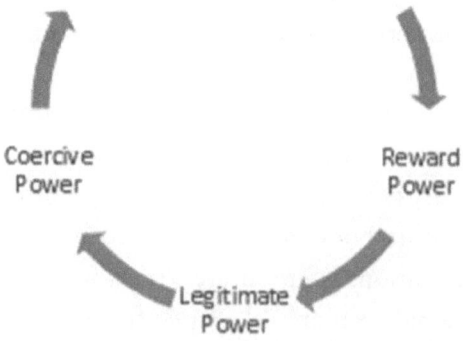

TRANSACTIONAL POWER BASES

Coercive Power

Reward Power

Legitimate Power

Now that we have found an ostensible home for three of the five power bases with regard to leadership style, what happens to the remaining two? Great question! Let's begin with expert power. We know from our previous discussion that expert power is that which is perceived by others to exist. There is no pretense involved. Others are not simply marching to the boss's drumbeat driven by position or hierarchy. Followers earnestly believe that you have knowledge or expertise deserving of admiration. Followers may also feel intellectually stimulated by your sharing of such knowledge. Likewise, as an expert, you may also be viewed as someone adept at innovation and problem solving. This expert view is very often associated with engineers. According to Bonasso (2001), "Engineers, as problem solvers are, by training, called on to consider the breadth and depth of the information they generate and the breadth and depth of the impacts of that information on the greater public system." These attributes (innovation and problem solving) are consistent with those observed in transformational leaders.

Leaders who are transformational are also admired. Followers seek to emulate their behaviors. Earlier we saw how referent power often stems from the desire of others to want to *be like* or *connect* with someone or some group. Recognizing that admiration could also occur in the negative context, the focus here is with regard to positive, ethical leadership. As a point of clarification, history reveals leaders admitted by some who behaved in such a way as to serve as a detriment to the better good. An example here would be those leading the Holocaust and everything that it represented. In this example, sure, there were those who connected with Adolf Hitler during this despicable moment in history. However, the desired end game was unscrupulous at best. Referent power, applied in an ethical context, is clearly linked to transformational leadership. There is also an implied trust component here. Followers who recognize that a leader is earnestly concerned with follower growth and development reciprocate with trust and exhibit increased levels of reference toward that leader. In the context of transformational leadership, leaders exhibiting growth and developmental concerns for their followers demonstrate idealized influence as well as individual consideration. Figure 3-3 reflects the two power bases linked to transformational leadership.

Figure 3-3: Transformational Leadership Power Bases

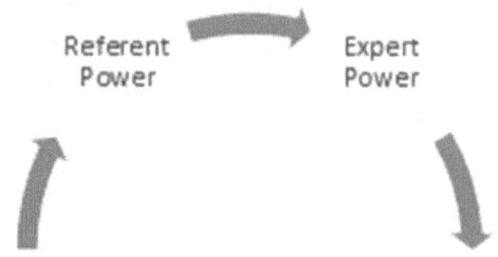

As indicated from the previous discussion, while legitimate, coercive, and reward power bases reflect mostly transactional events, expert and referent powers are closely aligned with transformational actions. But wait a minute. How can you be an effective leader without setting expectations, including goals and objectives, holding employees accountable and rewarding them if warranted? Great question! In summation, efficient and effective leadership requires full use of the five power bases. This point is further supported graphically by figures 3-2 and 3-3. You will note that neither set of power bases—transformational or transactional—closes the loop as a stand-alone. This is a very critical observation and is driven by the situational nature of leadership itself. The key here is recognizing when to apply each of the power bases driven by the particular situation at hand. As an example, it is quite necessary for a CEO to communicate the mission as well as expectations surrounding the accomplishment of organizational goals and objectives. At the same time, that CEO, acting in a more transformational manner, might focus on inspirational motivation versus dogmatic performance management as a means for achieving organizational progress.

At this point, it may be helpful to restate a bit of theory provided in the previous chapter regarding how leadership pioneers Avolio and Bass viewed the transformational and transactional relationships. According to Avolio and Bass (2004), some of the qualities associated with transactional leadership include "provides assistance in exchange for efforts, discusses who is responsible for what, makes clear [the] rewards for efforts, focuses attention on mistakes and attention [is] directed to failure." Transformational leadership includes "inspire, instill pride, sense of purpose, displays confidence, talks optimistically, articulates a vision [and] questions assumptions." From these comments, it should be apparent that effective leaders must both provide what is to be done and concurrently offer vision and strategies regarding how such may be accomplished. Supporting this point, Avolio and Bass (2004) stated that "the transactional process [contingent reward], in which the leader clarifies what the associates need to do for a reward, is nevertheless viewed ... as an essential component of ... effective leadership." Second, contingent reward—which may also be viewed as a transactional leadership power attribute—is only one of two transactional leadership constituents, thereby accounting for 50 percent of the total perceived style rating. The other constituent for transactional leadership is management-by-exception: active (see MBEA in appendix E). Another author (Bennett, 2009) cited works of multiple authors who argued that contingent reward is in itself related to transformational leadership. I think you get the point here. Both styles and all power bases are required for optimal leader effectiveness.

Despite the perception of my expert power in certain areas, the one thing that appeared time and again throughout my career was the need to refine existing skills and to develop new ones. Indeed, only through continuous improvement was I able to adapt to new circumstances and welcome challenges.

Having laid a comprehensive foundation for leadership, leadership styles, measurement constructs, and power, the next chapter lays out the critical parameters of my doctoral research regarding the influence of engineering education on perceived leadership style.

CHAPTER 4

Critical Research Parameters

SEVERAL QUESTIONS COME to mind when considering engineering education. McCuen (1999) captured some of the most critical ones as follows: "Does engineering education in its traditional form produce entry level engineers who have the necessary knowledge and abilities of a potential leader? Does engineering education even encourage its students to appreciate the importance of leadership potential? Does engineering education instill the belief that the ability to apply technical skills is what will lead them to success?"

The intent of this quantitative methods study was to determine the relationship—if one exists—between engineering education and leadership style. The independent variable, specifically engineering education, is defined by certified project managers (CPMs) and non-CPMs with engineering degrees and the same without engineering degrees. Thus, engineering degrees are expected to serve as a surrogate for engineering education. Predicated on the previous theoretical discussions, the dependent variable, namely leadership style, is defined as transformational and transactional. The interval-based Multifactor Leadership Questionnaire (MLQ) (see appendix B for partial sample form) was employed to assess the presence of the dependent variable among the targeted population. Although doing so was beyond the scope of this research, results from this study may serve as the impetus for further research aimed at addressing the broader question of whether or not a relationship exists between engineering skills and effective leadership.

4.1: Problem

Many Fortune 500 companies employ specific programs aimed at developing the core skills and business acumen for future organizational leaders. Such programs are typically referred to as leadership development programs (LDPs). Oftentimes these same companies employ multiple LDPs. General Electric, for example, offers LDPs in the areas of communications, finance, information technology, manufacturing operations, sales, and marketing. Each LDP necessitates dedicated infrastructure for its respective execution, which in turn requires resource allocation that is often redundant. If multiple LDPs could be supplanted with one LDP, leveraging highly talented entrants, economic benefits would be realized through reduced infrastructure for the support of multiple programs. Intuitively, engineers are potentially an excellent feeder pool for such a replacement program as they are tremendous thinkers and, given the rigor of their curriculum, have demonstrated resolve in the face of complex problems and challenges—not that I'm at all biased.

Thus, the author's aim was to determine the role—if any—that engineering education plays in perceived leadership style as exhibited by CPMs and non-CPMs holding engineering degrees (e.g., EE, ME, IE, etc.) versus the same without engineering degrees. A secondary goal was to determine which style within the management category (e.g., transformational or transactional) serves as the dominant style of leadership. With this in mind, the independent variable, CPMs with and without engineering degrees, is operationally defined consistent with Project Management Institute's Project Manager Professional (PMP) certification as documented per the following web address: http://www.pmi.org/en/Certification/Project-Management-Professional-PMP.aspx. Non-CPM managers are operationally defined as those with and without engineering degrees from whom direct reports or matrix-level reports receive their day-to-day work assignments. The integration of these groups would be operationally defined simply as the integrated population with and without engineering degrees. Leadership style, the dependent variable, is operationally defined by the transformational leadership and (TL) and transactional leadership (XL) constructs consistent with discussion in chapters 2 and 3. As mentioned, TL includes influence and motivation while XL focuses on rewards and punishment avoidance.

4.2: Method and Procedure

Employing a quantitative approach, the proposed research sought to assess leadership styles as a function of engineering education. To minimize noise associated with this proposal, the original approach was to measure leadership styles among several populations—CPMs, non-CPM managers, the integrated population with engineering degrees, and the same without engineering degrees. Thus, the critical research questions are as follows.

1. Does the integrated population with engineering degrees exhibit a leadership style (e.g., TL or XL) that statistically differs from the leadership style of the integrated population without engineering degrees?
2. Does a predominant style of leadership emerge when comparing within the two populations (managers with and without engineering degrees), and if so, what is it?

Although the two research questions may appear to be somewhat redundant, there is a subtle yet significant difference. In item 1, the object is to determine differences when comparing two distinct groups to each other (i.e., those with engineering degrees versus those without). The focus of item 2 is to look within each of the two groups and determine which style is more prevalent. Whether or not managers with and without engineering degrees are statistically different in leadership style than one another, is there a preferred style within each group.

The H1 hypothesis associated with this study stated the following: There is no statistically significant difference between leadership styles of managers (CPMs, non-CPM managers, or the integrated manager group) with engineering degrees versus the same without engineering degrees. The H2 hypothesis stated the following: No predominant style of leadership is evident among CPMs, non-CPM managers, or the integrated population with or without engineering degrees. In an effort to address hypotheses H1 and H2, sample population descriptive statistics were formulated and tested by employing parametric statistical approaches. In particular, the independent sample's t-test was used for the comparison of population means for perceived leadership style scores, analysis of variance (ANOVA) was employed to test multiple comparisons of the perceived mean of leadership style scores, and the one sample t-test was used to test the perceived mean of leadership style scores versus a gold standard.

The research environment was the domestic manufacturing environment facilitated by the internet. Leveraging Survey Monkey, an online survey resource, the MLQ (see appendix B for partial sample form supplied by Mind Garden) was issued to raters who reported directly or on a matrix basis to managers as described previously. Approval to conduct this human subject research was approved by the Old Dominion University Internal Review Board (ODU IRB). The survey process, which is available in appendix F, required the submission of participant profile information to the survey hosting company Survey Monkey. The hosting company then selected the participants based on the profile data provided. Based on discussions with the hosting company, it was believed that the greatest opportunity for yielding the desired sample population was to solicit participant responses from the manufacturing industry. Participants were directed to the Survey Monkey site and given the option to participate in the survey or exit the survey. The completed surveys were retrieved by the researcher for compilation and analysis. The experiment utilized a five-point Likert-type scale-based questionnaire as shown in appendix B to capture leadership style results for each of the manager categories with engineering degrees and the same without engineering degrees. In order to collect applicable demographic information, the questions shown in the box on the sample form in appendix B were modified by the researcher and posed to the subjects as presented in this section.

Additionally, operating within the spirit of the ODU IRB approval, all questions that would otherwise allow for the identification of the participant or the individual being rated by the participant were removed from the modified and issued MLQ. Shown in figure 4.2-1 are the specific questions, which were designed to first document the voluntary participation in the research and to subsequently collect demographic information for use in future research. As a side note, each of the questions asked of the participants, although recreated here, were actually presented to the participants in electronic form. Of course, that means that no answers were write-ins. They were all electronically selected.

Figure 4.2-1: Researcher Questions 1–3

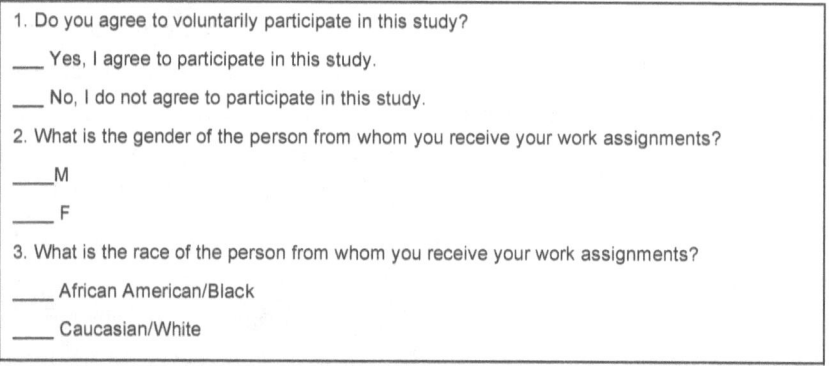

Referring to figure 4.2-2, question 4 sought to collect additional demographic data, while question 5 was intended to buttress responses to subsequent researcher questions.

Figure 4.2-2: Researcher Questions 4 and 5

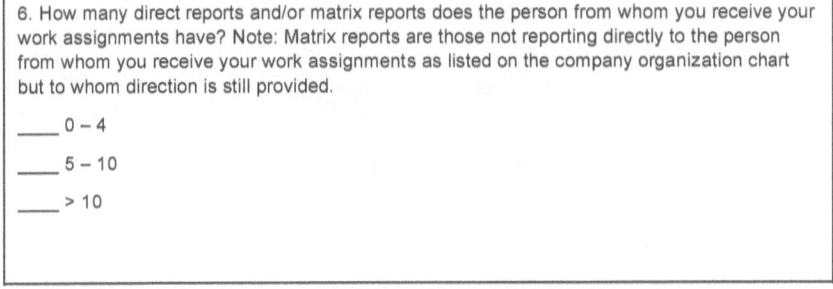

The aim of researcher question 6 (see figure 4.2-3) was to establish the span of control for the individual being rated.

Figure 4.2-3: Researcher Question 6

6. How many direct reports and/or matrix reports does the person from whom you receive your work assignments have? Note: Matrix reports are those not reporting directly to the person from whom you receive your work assignments as listed on the company organization chart but to whom direction is still provided.

____ 0 – 4
____ 5 – 10
____ > 10

Questions 7 and 8 (see figure 4.2-4) were the most critical to the current research. The initial research approach was to establish a sample population of CPMs with and without engineering degrees. Thus, in question 7, the rater was asked to identify the PMI certification status. The aim of question 8 was to establish whether or not the individual being rated possessed an engineering degree. In both questions, the option "not sure" was introduced to assure the highest possible integrity of the responses.

Figure 4.2-4: Researcher Question 7 and 8

> 7. Is the person from whom you receive your work assignments a certified project manager as evidenced by holding a certification received from the Project Management Institute (PMI)?
>
> ____ yes
>
> ____ no
>
> 8. Does the person from whom you receive your work assignments have a bachelor's, master's or higher-level degree in engineering (e.g. electrical, mechanical, industrial, chemical, etc.)?
>
> ____ yes
>
> ____ no

Questions 9 and 10 (see figure 4.2-5) were included to assist in determining possible areas for future research.

Figure 4.2-5: Researcher Question 9 and 10

> 9. Approximately how many years of professional work experience do you have?
>
> ____ Less than 4 years.
>
> ____ Greater than 4 but less than 10 years.
>
> ____ 10 or more years.
>
> 10. Do you possess a bachelor's, master's or high-level degree?
>
> ____ Yes
>
> ____ No

CHAPTER 5

Introduction to the Research

I WOULD BE REMISS if I didn't warn you that this and the next couple of chapters are steeped in academic and research jargon. Thus, if you have an affinity for the underlying detail for the research, you will by no means be underwhelmed. If, on the other hand, you prefer reviewing the research results and conclusions only to refer to the detail as needed, then feel free to skip over to chapter 8. Aimed at facilitating your ability to link leadership theory to the presented research, certain comments will be included verbatim in this chapter as were covered in previous chapters.

Okay, here we go! The notion of transformational and transactional leadership being theory is not as commonly accepted as, for example, the theory regarding relativity. Indeed, (Barling et al., 2010) stated of this perspective regarding TL, "We use the word theory because it is most familiar to practitioners, but we acknowledge that much of the new research ... would not fit that work in its strictest definition." Notwithstanding arguments regarding the application of theory in this context, literature addressing the theoretical foundation of leadership abounds.

Effective leadership is not simply about implementing canned tools, understanding models, or applying the traditional carrot-and-stick rules. Despite the approach employed, any effective attempt at leadership must take into account the psychology of leadership—the cognitive process. I repeat comments offered in chapter 3 once more here. Avolio and Bass (2004) reported that "when all levels of managers, students, and project leaders around the world were asked to describe the characteristics and behaviors of the most effective leaders with whom they had worked in the past," the characterizations were more transformational than transactional. Among the specific descriptors used for these leaders were "inspirational, intellectually stimulating, challenging, visionary, development oriented, and determined to maximize performance." These characterizations essentially mirror the five constituent elements of TL.

Precedent for assessing the presence of transformational and transactional leadership attributes in the general area of leaders and followers (e.g., project teams) is provided by Hoyt and Ciulla (2004), as cited by Brooks, Levine, and Muenchen (2010), with the following comments: "Transformational leadership … examines the relationship between the leader and the followers and focuses on issues relating vision, risk-taking, enthusiasm and confidence." Similar sentiments exist regarding the XL style as it, too, assumes a leader-follower environment for its execution.

5.1: Literature Review

The initial focus of this research was to assess leadership style influence based on a test group comprising certified project managers (CPMs). Aimed at facilitating research toward addressing the previously outlined purpose statement, a literature tree was developed and implemented. As indicated in figure 5.1, the first step in addressing the main problem was to determine appropriate categories that may afford statistical comparisons between representative groups of candidates possessing engineering education and those without such education. The author searched literature databases for books, journals, and other materials in the area of engineering-oriented leadership categories. Here, the author sought to identify such categories that are commonly recognized and positioned in a leadership hierarchy. In an effort to further minimize potential noise, incumbent criteria (e.g., project management certification) were established. Thus, CPMs with and without engineering degrees served as one comparison set within the independent variable. Next a similar search of literature databases (e.g., books, journals, etc.) was conducted in the area of leadership (see subproblem 2). The aim was to identify references to leadership theory that are commonly understood to be both observable and quantifiable. Thus, transformational and transactional leadership styles (comprising the dependent variable) were selected for assessment when considering the target population—CPMs and non-CPM managers with and without engineering degrees. Leveraging the literature reviews, the final step in this area (see subproblem 3) was to identify Likert-type scale-based survey instruments that were commonly regarded as validated per scholarly and peer-reviewed writings. As will be discussed in subsequent sections, the MLQ was employed for leadership style assessment.

Figure 5-1: Literature Review Tree for Proposed Research

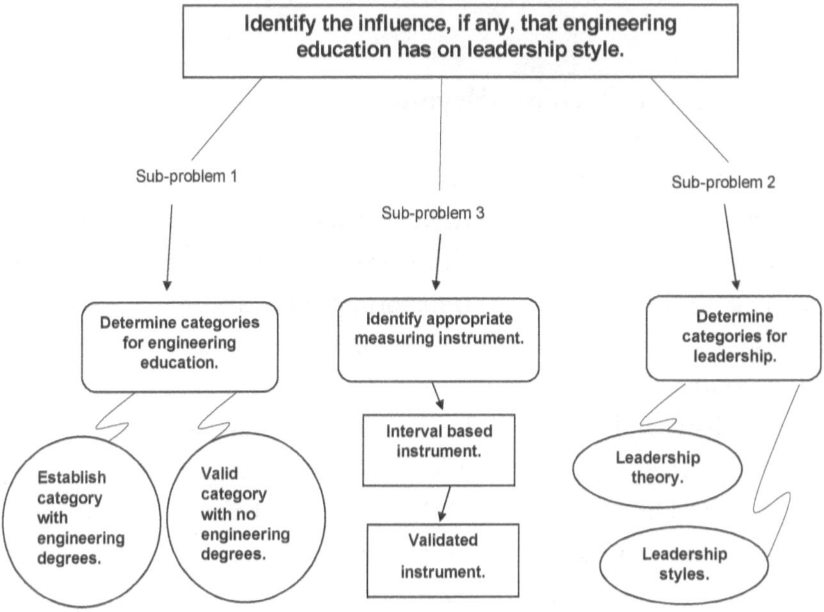

```
                Identify the influence, if any, that engineering
                education has on leadership style.

        Sub-problem 1          Sub-problem 3          Sub-problem 2

        Determine categories   Identify appropriate   Determine
        for engineering        measuring instrument.  categories for
        education.                                    leadership.

                               Interval based
                               instrument.           Leadership
    Establish      Valid                             theory.
    category       category
    with           with no     Validated
    engineering    engineering instrument.           Leadership
    degrees.       degrees.                          styles.
```

5.2: Literature Review Detail

While literature reviews at the time of this writing provided much insight regarding the areas of project management and leadership style, such reviews did not identify a study or studies assessing the extent that engineering training may or may not influence leadership style. From the literature review, it was also clear that while much research had been implemented in the area, little to no categorization of the raters or those being rated was identified. Thus, it was not known, for example, whether or not the project managers were certified. The levels and/or types of education documented were not known either. As a point of fact, less than 50 percent of the reviewed articles demonstrated evidence of project manager assessment for TL, and of those (see appendix A), two-thirds either only assessed one factor, were gender-biased, or did not specifically point to the subject being assessed as the project manager. Likewise, while assessments of TL were present in all articles retrieved, close to 80 percent of them either referenced other studies, only assessed one factor, assessed portfolio managers, or was gender-biased.

5.3: Literature Review—Beyond the Gap

Notwithstanding the lack of categorization of the sample group as discussed previously, the reviewed literature did offer insights as to the potential for linkages between project management and various leadership styles, including transformational and/or transactional leadership (Deanne, Hartog, and Keegan, 2004; Ryoma and Tapanainen, 1999; Neuhauser, 2007; Muller and Turner, 2010; Prabhakar, 2005; Kissi, Dainty, and Tuuli, 2013). Attempting to assess the presence of TL within the leader subordinate group is quite prevalent. Indeed, according to Deane and others (2004), "Transformational leadership is a concept that has come to prominence in the last two decades." And, while Deane and others (2004) hypothesized that "transformational leadership style is positively related to employee commitment and negatively to employee's perceived stressfulness of the job" Muller and Turner (2010) found that "transformational leadership, and concern for people, is necessary on more-demanding projects." Although the former hypothesized relationship was not supported by study results, taken in concert with findings by Muller and Turner (2010), such might suggest that the TL style becomes even more important as project demands increase, particularly with regard to motivation, which is a key component of the TL style and which is also respectively measured by the MLQ instrument (Schriesheim et al. 2009). Notwithstanding Muller and Turner (2010) hypothesizing that "the project manager's leadership style influences project success and that different leadership styles are appropriate for different types of projects" they also cite studies suggesting that motivation falls under the emotional competency category, advancing a slightly different leadership style construct. In this vein, Dulewicz and Higgs (2003), as cited by Muller and Turner (2010), "identified fifteen [competencies] which influence leadership performance." They group the competencies into three competence types, which they term "intellectual (IQ), managerial (MQ) and emotional (EQ)." Neuhauser (2007) also cited Rosener (1990) who "found that women tend to use transformational leadership more than men." However, as also pointed out by Neuhauser (2007), "The behaviors identified as the most important (absolutely vital and important) [in project leadership included attributes of] transformational, [and] transactional [leadership]." As previously referenced, one of the key aspects of TL is providing a clear vision. Indeed, according to Lussier and Achua (2009), as cited by Spieth, Tyssen, and Wald (2013), "A transformational leader focuses on people and their motivations, beliefs, and behaviors, and provides them with visions that satisfy their needs and desires." Similarly, Christenson and Walker (2004) concurred, arguing that "a significant driver of project management success is effective and intelligent leadership communicated through an inspiring vision of what the project is meant to achieve and how it can make a significant positive impact." And insomuch as vision is subsumed by the TL style, it could be argued that TL is therefore instrumental in project success. To this end, Kissi and others (2013) hypothesized that "transformational leadership behavior of portfolio managers positively influences project performance." Study results offered support for this hypothesis as follows: "Transformational leadership [has] a significant and positive relationship with project performance ($\beta = 0.328$, $\rho = 0.001$) and explains 10% of the variance in project performance."

Thus, the linkage exists between TL and project success. Andreas and others (2013) went a step further in defining the linkage between project leadership and project performance, hypothesizing that "transformational leadership behavior will be especially effective in projects that have strong goal clarity rather than path-goal uncertainty [and that such leadership] will be especially effective in short project durations." However, despite apparent literature-based support offered for the propositions, the authors also stated, "We have not yet empirically confirmed these findings ... we thus advocate the empirical testing of our propositions."

To this point, the literature review has focused on the project manager for the most part, and as such, assumptions have been made regarding the team. Haung and others (2011) built on this thinking, specifically addressing the relationship between leadership style, teamwork, and project success. They hypothesized that "the project manager's leadership and teamwork (in terms of team communication, collaboration and cohesiveness) are correlated [that] teamwork (including team communication, collaboration, and cohesiveness) and overall project success are correlated [and that] project type may act as a moderator between teamwork and overall project success." Reflecting on these hypotheses, it would appear logical that the more unified the team, the greater the chances are for success. Accordingly, Haung and others (2011), referring to their study results, found that "the results from this analysis suggest that all three composite measures (project manger's leadership, teamwork, and overall project performance), are highly correlated." Thus, these comments might suggest that while leadership style plays a very critical role in project success, it is the team's reception of that style that serves as the impetus for such success. Building on the importance of the team's view and acceptance of project manager leadership style, Bennet (2009) hypothesized that "there is a relationship between the subordinate's perception of leadership style of IT managers and the subordinate's perception of IT managers to inspire extra effort." Not only was a strong correlation identified between TL and extra effort, but the study also found that "transformational leadership subscales of idealized influence (attributed), idealized influence (behavior), inspirational motivation, intellectual stimulation and individualized consideration also had strong correlations." Despite that apparent and logical linkage between leadership style and project success and as a sort of moderating factor, Muller and Turner (2005) "were commissioned by the Project Management Institute to determine: 1) whether the competence, including personality and leadership style, of the project manager is a success factor for projects; and 2) if different competence profiles are appropriate for different project types." Among the study findings were the following comments: "It is conceivable that the leadership style and competence of the project manager have no impact on project success, and the unique, novel, and transient nature of projects (as well as the risk involved) means the leader has less of an impact on performance. But that question can only be answered if it is directly measured." Thus, while the preponderance of the literature reviewed does point to a potential linkage between TL and project leadership, it does not provide clear categorical evidence regarding the role that engineering education plays in leadership style exhibition. Nor does it address the project manager credentials. These two opportunity areas were therefore addressed by my doctoral research.

In the spirit of implementing research that contributes to the body of knowledge, an additional literature review was conducted considering the expanded sample population (managers without PMI certifications). I was successful in locating articles that focused on TL and the education sector as well as one article (supplied by Sibel, Olga, Alabart, and Medir, 2013) that focused on allowing fourth-year engineering students who were enrolled in a "project management course, the opportunity to develop their team leadership competencies." Another article authored by Collado, Laglera, and Montes (2013) conducted structural equation model testing to assess the "effects of leadership style on engineers." However, this Spain-based study did not assess the leadership style of the engineers themselves. Instead it assessed the effect of the engineers' superior's leadership style on the engineers' attitudes as subordinates. Despite considering the expanded sample population, a gap remained in the literature regarding the relationship—if any—between engineering education and leadership style.

CHAPTER 6

Methodology

OKAY, WE'RE ALMOST there, but first, we need to discuss the research approach. A quantitative method was employed. And as mentioned previously, Likeret-based survey instruments were used to collect data for population of the two category distributions (managers with and without engineering degrees). Survey Monkey, an online survey company, was enlisted to identify participants belonging to the categories of interest. The survey was "cross-sectional with the data collected at one point in time" as espoused by Creswell (2009). During the initial planning, the author considered various tools associated with the qualitative method (e.g., interviews) for collecting category data. And while interviews would certainly allow for researcher insertion and possibly add context to data collected, such an approach can be quite protracted as well as cost and time prohibitive. For example, assuming an "n = 100" per category, such would require as many one-on-one interviews. This also assumes, of course, that the author is able to secure time with each of the interviewees across varying corporate and possibly geographic environments. There is also the potential issue of noise inherent with in-person interviews. For example, were the author to conduct such interviews, interpretations of responses would be tied to the author's views of the world, which may not necessarily be aligned with the views of those being interviewed. This perspective is shared somewhat by Noonan (2013) in the following comments regarding the disadvantages of interviews: "The researcher's views can influence the participant's responses by expressing surprise or disapproval." Thus, the author elected to implement a purely quantitative methods approach in conducting the current research.

6.1: Literature Review-Based Research Paradigm

Figure 6.1-1 models the selected research paradigm for the current study to include the worldview or ontological position as well as the epistemological stance, the method employed, and the mode of reasoning selected. Creswell (2009) talked about the need for positivists to identify and assess causes that influence outcomes that aligns with the current research plan relative to engineering education and leadership style. This, of course, can only occur empirically through others' observation of leadership style as was the case employing the MLQ. Although some may argue the objectivity of results predicated on individual observation and suggest that such more closely aligns with a qualitative methodology, the fact is that the current research quantifies the survey results and begins to shift the plan to the quantitative method. Many scholarly writings support this perspective, including work by Leedy and Ormond (2013), who stated that "a quantitative researcher typically tries to measure variables in some numerical way [including] tests, questionnaires [and] rating scales." Creswell (2009) regards this approach as residing in the quantitative space as well, stating, "A survey design provides a quantitative or numeric description of trends, attitudes, or opinions of a population by studying a sample of that population." Likewise, Hathaway (1995) also sanctioned the use of surveys in quantitative research by stating, "A quantitative approach [includes] surveys and statistical analysis of responses [versus] qualitative approach (e.g., transcription analysis of interviews)."

Because the current research was not seeking to develop a theory of leadership but instead test an existing theory, the mode of reasoning was clearly deductive as opposed to inductive. Support for this position was offered by Popper (1992), as cited by Siangchokyoo and Sousa-Poza (2012), stating, "During the deductive research process, the researcher stipulates an idea (hypothesis), performs some form of experimentation, and collects data to verify if the results are consistent with the postulated hypothesis."

Figure 6.1-1: Research Paradigm

Ontological Position	Epistemological Position	Methodology Employed	Mode of Reasoning
Positivist Versus Constructivist	**Empiricist** Versus Rationalist	**Quantitative** Versus Qualitative	**Deductive** Versus Inductive

Consistent with the previously referenced positivist worldview, four basic rules, or cannons were also selected for the current research. First, there must be internal validity such that the author is able to draw accurate conclusions regarding any of the relationships presented in the data. Second, the study must also have good external validity, which in turn would allow the results to be generalizable to the broader context. Implementation of this second cannon is supported by Leedy and Ormond (2009), who stated, "Researchers contribute more to humanity's knowledge about the world when they conduct research that has implications that extend far beyond the situation being studied." Third, particularly as it relates to the measuring instrument, in this case a survey questionnaire that will be discussed in detail later, the research has to provide reliable results.

And finally, the research must provide objectivity. Thus, as mentioned previously, the positivist position was taken with the current research, resulting in the development of data consistent with mind-independent review and neutrality. Enabling the objectivity platform on which the positivist approach is founded is the objectivity of findings predicated on implementation of a measuring instrument with proven validity and reliability as will be discussed in the next section.

6.2: Measuring Instrument

Indeed, as previously discussed, there are many instruments (e.g., FFM) that may be used to measure leadership style. Critical to the instrument chosen, however, is its reliability and validity.

Leedy and Ormond (2013) offered support for this perspective by stating, "Regardless of the type of scale a measurement instrument involves, it must have both validity and reliability for its purpose." Likewise, "although individuals may have different views in terms of what constitutes psychometric adequacy, most people can agree that a measurement is only useful to the extent that it provides meaningful information about individuals" (Briesch, Chafouleas, and Swaminathan 2014). Creswell (2009) added to the discourse by stating, "To use an existing instrument [the author should] describe the established validity and reliability of the instrument [which includes] reporting efforts by authors to establish validity." Figure 6.2-1 reflects the interconnectedness of the relationships between and among the instrument of choice and the critical components of validity, reliability, and objectivity.

Figure 6.2-1: Instrument Reliability and Validity

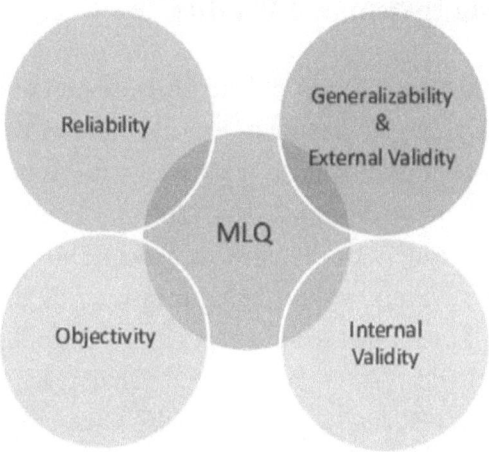

Prior to initiating discussions regarding instrument validity and reliability, it is appropriate to first visit the MLQ in the context of full-range leadership theory (FRLT) (refer to appendix E). Pioneering authors of leadership theory, such as Bass and Avolio, determined that more was needed than leaders simply providing rewards for subordinate behavior characterized by XL. They also identified the need to understand how leaders influence followers to set aside self-interests for the good of their organizations through optimal levels of performance. Early expansions in leadership theory included five TL factors, three XL factors, and one nontransactional laissez-faire leadership component (Antonakis et al.). The contemporary FRLT model maintains the five TL factors as discussed previously—idealized influence, idealized behaviors, inspirational motivation, intellectual stimulation, and individualized consideration. However, the XL factors total to two and are defined as contingent reward (CR) and management-by-exception: active (MBEA). The final leadership style, passive avoidant, is also comprised of two attributes—management-by-exception: passive (MBEP) and laissez-faire (LF). The MLQ (see appendix B) is designed to assess each of the three leadership styles through select questions that are subsequently combined via the MLQ5X form (see the partial form in appendix C) for determination of applicable descriptive statistics.

6.3: Measuring Instrument Validity

We begin with discussions of instrument external validity. Leedy and Ormond (2013) characterized external validity as "the extent to which the research study's results apply to situations beyond the study itself." According to Avolio and Bass (2004), "In numerous studies, transformational leaders were found to generate higher commitment in their followers." Thus, what is being measured by the MLQ can be traced to a valid form of real-world effective leadership. Likewise, testing conducted by Bogler (2001) determined "that teachers' satisfaction increases as they perceive their principals' leadership style as more transformational and less transactional." Fuller and others (1996), as cited by Avolio and Bass (2004), reported in a meta-analysis greater follower compliance if their leaders were more transformational than transactional (Avolio and Bass 2004). The list of scholarly writings substantiating the external validity of the MLQ is far-reaching. Thus, discussions in this section will shift to construct validity.

"The extent to which an instrument measures a characteristic that cannot be directly observed but assumed to exist based on patterns in people's behavior [is termed construct validity]" (Leedy and Ormond 2013). Creswell (2009) addressed the topic of construct validity by asking, "Do items measure hypothetical constructs or concepts?" According to Barge and Schlueter (1991), "The MLQ possesses good construct validity … as seen in the previous studies, transformational versus transactional leadership was found to be more highly correlated with a variety of outcomes."

Following the implementation of tests that included confirmatory factor analysis, modification indices, and chi-square testing of the nine factor model (i.e., the MLQ version used in the current research), Armstrong and Nuttawuth (2008) concluded, "After acknowledging the MLQ criticisms by refining several versions of the instruments, the version of the MLQ, Form 5X (Bass and Avolio, 1997), is successful in adequately capturing the full leadership factor constructs of transformational leadership theory." In the end, there appears to be significant support for the MLQ's construct validation.

Regarding predictive validity, according to Barge and Schlueter (1991), "The MLQ demonstrates good predictive validity. Bass and Avolio (1990) report transformational leadership scores were strongly correlated with the extra effort of followers, satisfaction, and the effectiveness of the organization."

6.4: Measurement Instrument Reliability

Leedy and Ormond (2013) defined reliability as "the consistency with which a measuring instrument yields a certain, consistent result when the entity being measured hasn't changed." Bass and Avolio (1991), as cited by Barge and Schlueter (1991), concluded that although "the alpha reliability coefficients for the self- rating form were consistently lower than those for the rater form, with reliabilities ranging from .60 to .9 [however] reliability of the two forms existed." (It should be noted that the current research utilizes the rater form for data collection.) Bass and Avolio (2004) concluded that "reliabilities for the total items and for each leadership factor scale ranged from .74 to .94 … all of the scales' reliabilities were generally high, exceeding standard cut-offs for internal consistency recommended in the literature." Barge and Schlueter (1991) also "report[ed] the MLQ Rater Form demonstrated good internal reliability with all factors above an alpha of .82, with the exception of management-by-exception (.79) and laissez-faire leadership." Also in this area, Bennett (2009) cited research conducted by Lowe, Kroech, and Sivasubramaniam (1996) that assessed five factors of the MLQ, including charisma, individualized consideration, intellectual stimulation, contingent reward, and management-by-exception. The resulting "mean Cronbach scale obtained for the five scales tested were 0.92, 0.88, 0.86, 8.82 and 0.65 respectively." Bennett (2009) also cited work by Dumdum, Lower, and Avolio (2002) that assessed "twelve scales" of the MLQ, concluding that "internal reliability was good as the mean Cronbach ... for eleven of the twelve scales was above 0.7 and the final one was 0.69."

There is another reliability measure termed test- retest reliability, and according to Bass and Avolio (1990), cited by Barge and Schlueter (1991), "test-retest reliabilities were provided by a study using the ratings by 193 followers and 33 leaders measured 6 months apart ... the rater form test-retest reliabilities ranged from .52 to .82 and from .44 to .74 for the self-rating form."

CHAPTER 7

Research Execution

WHEW, THAT WAS a lot of information. It is my hope thus far that I have not totally discouraged you, or anyone else, from aspiring to complete a doctorate of philosophy. As a reward for your patience and diligence, we have now reached the point that you have been waiting for—the results of my research.

As discussed in section 4.2, the original aim of the research was to specifically test the previously discussed hypotheses with respect to CPMs alone. Thus, the sample population was to only include CPMs with engineering degrees and those without engineering degrees. In practice, however, economic constraints limited the total sample population of this category to sixty-seven (forty-eight with engineering degrees, 15 without and 4 unusable). Employing SPSS Sample Power 3, based on the pilot testing for CPMs, 113 subjects per group (with and without engineering degrees, totaling 226) would be required to yield a power of 80 percent, and of the 350 completed surveys received, CPMs meeting the desired criteria accounted for only sixty-three (18 percent) of the total number of rated organizational leaders. Consequently, while all testing included the CPM group, the scope was expanded to also include managers with and without PMI certifications as well as managers with and without engineering degrees. However, all managers were responsible for providing the day-to-day work activities for one or more reports (direct or matrix). As previously discussed, this population, inclusive of CPMs, was termed the "integrated population." Based on this population pilot testing, 116 subjects per group (with and without engineering degrees, thus totaling 232) would yield a power of 90 percent. Expanding the scope of the current research to include the integrated population not only increased the statistical power of the testing because of increased cases available but also remained true to the fundamental research goal to determine the relationship—if any—between engineering education and leadership style by assessing the leader styles of those with engineering degrees and those without.

7.1: Population and Demographics

Those employed in the domestic manufacturing sector comprised the sample group. Based on information supplied by Survey Monkey, roughly five hundred prospective participants visited the site for potential survey completion. Of those, only 350 actually completed the survey. (This should not suggest a 70 percent conversion rate as it is not known to the researcher how many individuals were actually asked to complete the survey and elected not follow the link to the survey.) Figure 7.1-1 reflects the number of total cases (completed surveys) received as well as the group allocation for those cases. As previously indicated, of the 350 cases, close to 20 percent were not usable because of the "not sure" response provided by the raters under the questions regarding engineering education. Consequently, only 283 cases were potentially usable for testing the research hypotheses.

Figure 7.1-1: Database

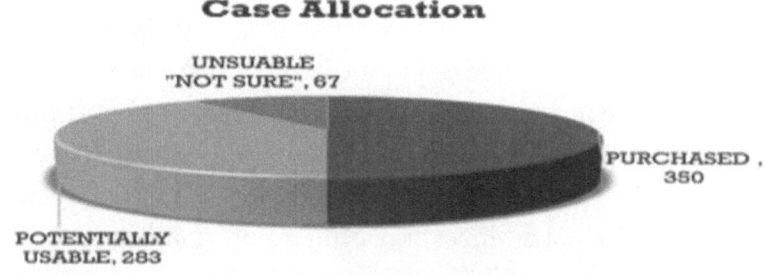

Results of the researcher question regarding engineering education are shown in Figure 7.1-2. Accordingly, the terms "no engineers" and "engineers" refer to whether or not the individual being rated possessed an engineering degree. As shown in figure 7.1-2, sixty-four CPMS were reported to possess engineering degrees, while fifteen did not. For managers (MGRS) it was sixty-four with these degrees and 112 without them, and among the combined groups of CPMS and MGRS, referred to as the integrated group (INTGR), 111 were reported to have engineering degrees, while 169 did not. Group totals for CPMs, MRGS, and INTGR were sixty-three, 218, and 281 respectively.

Figure 7.1-2: Education Data

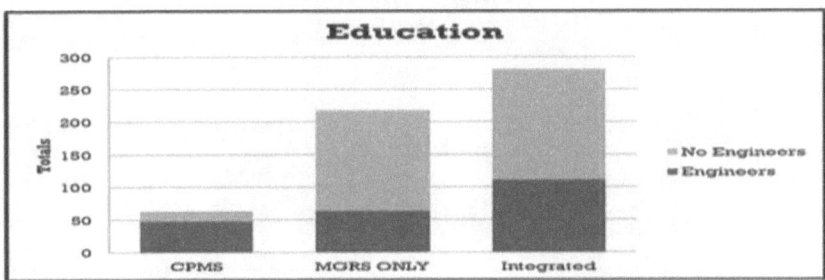

Because the largest sample population was in the integrated population, demographic data will be reviewed in that context. The first bit of demographic data has to do with gender as shown in Figure 7.1-3. Among the integrated population, there were thirty-four females with engineering degrees and seventy-six without. For males, it was seventy-six and ninety-one respectively.

Figure 7.1-3: Gender

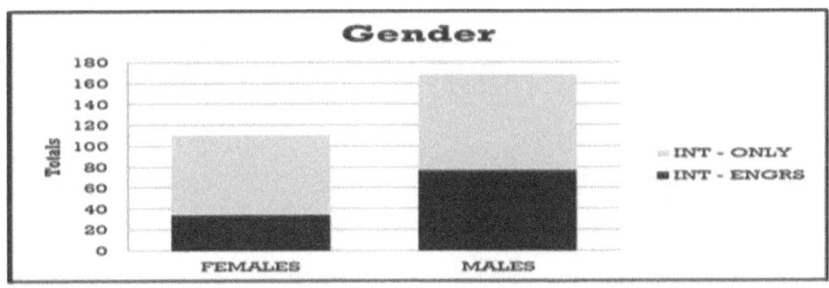

Figure 7.1-4 reflects the experience level of the individuals being rated. Among the integrated population with engineering degrees (INT ENGRS), there were fourteen with zero to four years of experience, thirty-seven with four to ten years of experience, and sixty with more than ten years of experience. The same data for those without engineering degrees (INT ONLY) revealed sixty-one, thirty-three, and seventy-four respectively.

Figure 7.1-4: Experience Level for Those Rated

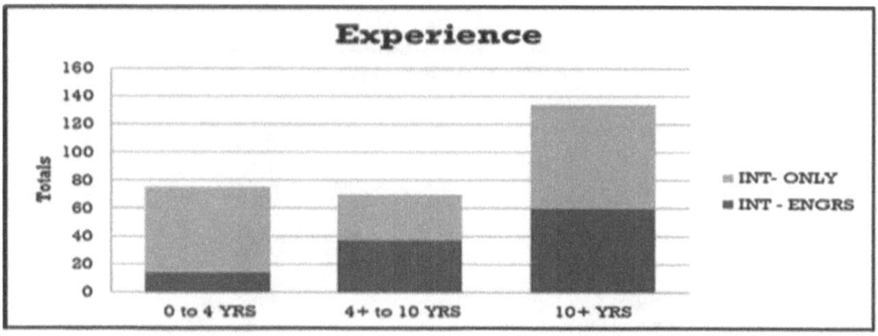

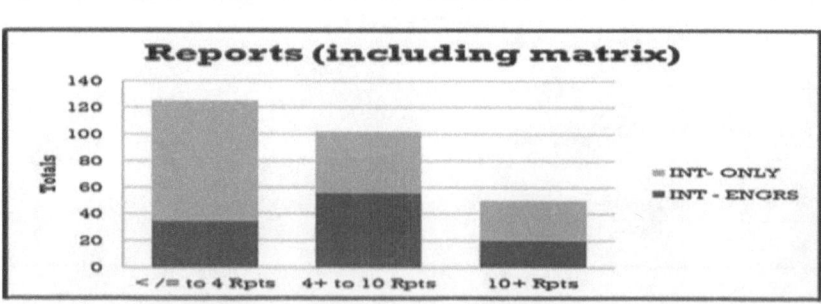

Figure 7.1-5 reflects the number of persons for whom the individual has day-to-day work assignment responsibility. Accordingly, among the group with engineering degrees, there were thirty-five with zero to four organizational reports, fifty-six with four to ten organizational reports, and twenty with more than ten organizational reports. The same for the group without engineering degrees revealed ninety, forty-six, and thirty respectively.

Figure 7.1-5: Organizational Reports (Direct or Matrix)

7.2: Data Analysis—Engineering Education and (TL, XL, PA)

Recall that TL = transformational leadership, XL = transactional leadership and PA = passive avoidant leadership. The SPSS statistical software package was used to facilitate data analysis for assessing the extent to which engineering education influences perceived leadership style (TL, XL, and PA). The first research hypothesis (H0a) asserted the following: There is no relationship between engineering education and transformational leadership (TL). This was restated to accommodate the appropriate statistical testing. Restated, we have the following: Ho: $\mu TLW = \mu TLWO$. This says that the population means for TL styles of CPMs with and without engineering degrees are the same. We also have the following: Ha: $\mu TLW \neq \mu TLWO$. This says that the population means for TL styles of CPMs with and without engineering degrees are different. Accordingly, this first test focused on the CPM groups with and without engineering degrees. Given that the groups are independent, the independent samples t-test was implemented. One of the assumptions that should be tested before applying the t-test is an assessment of the data in search of outliers.

Following the initial box plot run, two outliers (lines 42 and 23, which are not shown) were identified. In looking at the data, it appeared that a couple of the survey participants sort of flat lined the survey, entering a zero (not at all observed) for at least twenty-two of forty-five MLQ questions for CPMs with engineering degrees for line 42. Likewise, for outlier 23, a similar pattern was observed. Both outliers were eliminated. The box plot was then rerun, revealing another outlier (line 15) that was also removed for the same reason, yielding the box plot shown in Figure 7.2-1.

Figure 7.2-1: Adjusted Box Plot

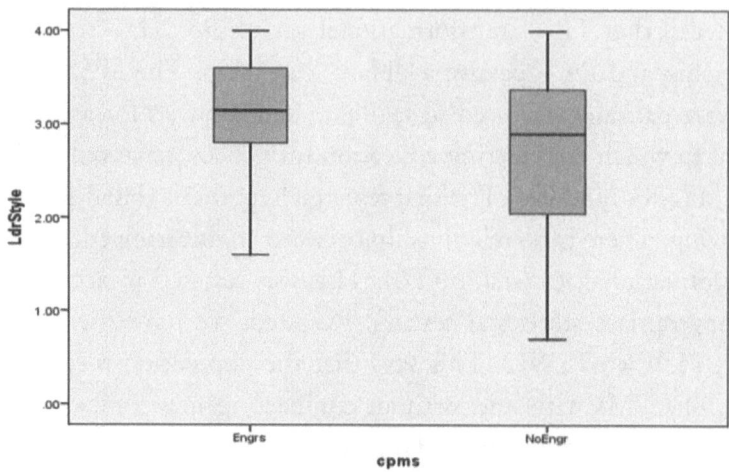

Table 7.2-2 reflects the results from the test for normality. As indicated, the Shapiro-Wilk test generated P values (located in the Sig columns) suggesting that while the distribution for the CPMs with no engineering degrees is normal, given by the Sig of 0.480, the same for CPMs with engineering degrees was only 0.011, thus suggesting an abnormal distribution.

Table 7.2-2: CPM Normality Test

	cpms	Kolmogorov-Smirnov[a]			Shapiro-Wilk		
		Statistic	df	Sig.	Statistic	df	Sig.
LdrStyle	Engrs	.107	45	.200[*]	.932	45	.011
	NoEngr	.122	15	.200[*]	.947	15	.480

*. This is a lower bound of the true significance.
a. Lilliefors Significance Correction

Don't worry about the usage of the t-test and the consequences of outliers or failed tests for normality. Elliott and Woodward (2007) cite "rules of thumb" offered by Moore and McCabe (2006), among which are the following: "If the sample size is large (at least 40), then the one-sample t-test can be safely used without regard to skewness or outliers." Although the current study also leverages the two sample t-test and ANOVA, Elliot and Woodward refers the reader back to these guidelines for both of these tests as well. From the t-test (see table 7.2-3), the significance value is .026. Consequently, the assumption of homogeneity of variances was not met. Thus, the "Equal variances not assumed" row was used for decision making. Because the t-test at t(18) degrees of freedom returned "Sig" or p = .164, which is greater than .05, it cannot be concluded that a statistically significant difference exists between the two perceived TL-style mean scores for CPMs with and without engineering degrees. Consequently, possibly driven by the low power of the test, which was less than 80 percent, the null hypothesis cannot be rejected.

Table 7.2-3: CPM Independent Samples t-Test

		Equality of Variances		t-test for Equality of Means					Interval of the	
		F	Sig.	t	df	Sig. (2-tailed)	Mean Difference	Std. Error Difference	Lower	Upper
LdrStyle	Equal variances assumed	5.199	.026	1.797	58	.078	.39889	.22200	-.04549	.84327
	Equal variances not assumed			1.451	18.037	.164	.39889	.27496	-.17870	.97647

Although the first research hypothesis was limited to TL, given the availability of information offered by the MLQ regarding the full range of leadership, the same tests were implemented for transactional and passive avoidant (PA) leadership styles. Again, restating the hypothesis to accommodate this test, we have the following: Ho: µLdrstylei = µLdrstylej. And conversely, we have the following: Ha: µLdrstylei ≠ µLdrstylej. In the lexical sense, the general restated hypothesis is that there is no difference in leader style with I and J serving as a surrogate for the respective styles with and without engineering degrees.

Summarizing the analysis, because of unequal variances for XL, the "Equal Variances not Assumed" column was once again used for statistical decision making, and despite degreed CPMs having a perceived mean score of 0.48 higher, because t(18) degrees of freedom returned a "Sig" or p = .083, it cannot be concluded that a statistically significant difference exists between the two XL-style means for CPMs with and without engineering degrees. Thus, the null hypothesis cannot be rejected. Similarly for PA, although the degreed CPM's perceived mean PA leadership style was 0.41 higher than the PA leadership style for CPMs without degrees, because at t(60) degrees of freedom "Sig" = .180, it cannot be concluded that a statistically significant difference exists between the two PA-style means. Once again, the null hypothesis cannot be rejected. In an effort to simplify the presentation of statistical testing, table 7.2-4 (Composite Summary Statistics) lists the results of each test (normality, equality of variances, and significance testing) and also report the mean difference between the respective groups with engineering degrees in comparison to those without. And the results also reflect testing of all styles (TL, XL, and PA). For example, referring to the managers group and the TL style, it is evident that the normality assumption was not met for either distribution but that the equality of variance assumption was met. It is also evident that the mean difference between the TL style of the two groups was 0.47 with a confidence interval of 0.18 to 0.76 and that the sample sizes were N1 = 64 and N2 = 154 for managers with engineering degrees and the same without respectively. Finally, it is evident that T at 216 degrees of freedom was 3.24, and with a two-tailed P value of 0.001, the means were statistically different. You will note the slight differences in populations for some of the tested styles. For example, while there were sixty-four managers tested for TL (N1), only sixty-three were tested for XL. This is not an oversight and simply reflects that lack of survey responses for the specific constituent.

Table 7.2-4: Composite Summary Statistics

Style Tested	Group	Normality (p < 0.05)		Equality of Variance (p < 0.05)	Mean + CI		T-test Result		
		Engrs N1	NoEngrs N2		N1	N2	t	df	2 tail Sig
TL	MGRS	0.007	0.001	0.063	0.47 (0.18, 0.76) 64	154	3.24	216	0.001
XL	MGRS	0.786	0.007	0.006	0.43 (0.22, 0.64) 63	154	4.06	173	0.0005
PA	MGRS	0.165	0.006	0.296	0.10 (-0.15, 0.35) 64	149	0.79	211	0.432
TL	INTGR	0.0005	0.0005	0.001	0.63 (0.42, 0.85) 110	169	5.72	270	0.005
XL	INTGR	0.293	0.007	0.006	0.55 (0.36, 0.74) 111	169	5.68	273	0.005
PA	INTGR	0.038	0.012	0.046	0.15 (-0.07, 0.37) 112	162	1.37	217	0.172

☐ Indicates statistically significant differences

Summarizing the CPM testing and table 7.2-4 findings, no statistically significant differences were found in the CPM groups with or without engineering degrees for TL. However, in both the manager and integrated population, TL and XL were statistically different (two-tailed Sig less than 0.05) and higher for those with engineering degrees versus those without. Likewise, no statistically significant differences were found in the CPM groups with or without engineering degrees for XL, yet in both the manager and integrated population, TL and XL were statistically different and higher for those with engineering degrees versus those without. No statistically significant differences were detected in any of the groups for PA.

Returning once again to TL theory, recall that intellectual stimulation (IS) is one of its five constituent elements. And according to Avolio and Bass (2004), leaders demonstrating this attribute stimulate innovation and creativity by questioning assumptions, reframing problems, and approaching old situations in new ways. They also solicit new solutions to problems and include followers in the problem-solving process.

Considering the academic lessons learned by engineers, especially in the area of problem solving, the current research also considered whether or not statistically significant differences existed in the perceived demonstration of the IS attribute when comparing those with engineering degrees to those without. These results are summarized in table 7.2-5. Restating the hypothesis to accommodate this test, we have the following: Ho: μIS = μIS. The population means for IS in groups with and without engineering degrees are the same. And conversely, we have the following: Ha: μIS ≠ μIS. The population means for IS in groups with and without engineering degrees are different.

Table 7.2-5: Summary Statistics for IS

		Normality		Equality of Variance	Mean Diff + CI		T-test Result		
		(p > 0.05)		(p > 0.05)	N1	N2			Diff if p < 0.05
Style Tested	Group	Engrs N1	NoEngrs N2				t	df	2 tail Sig
IS	CPMS	0.006	0.44	0.425	0.234 (-0.26, 0.73) 46	14	0.951	58	0.345
IS	MGRS	0.172	0.002	0.016	0.56 (0.302, 0.83) 62	154	4.2	142	0.0005
IS	INTGR	0.003	0.002	0.0005	0.74 (0.53, 0.94) 121	169	6.9	288	0.0005

☐ Indicates statistically significant differences

From table 7.2-5, it is evident, as might be expected because of the low sample size, the null hypothesis stating that there is no difference between the mean perceived IS styles for CPMs with and without engineering degrees cannot be rejected. However, for the group of managers and the integrated population, the null hypothesis can be rejected, suggesting that those with engineering degrees may be perceived to demonstrate more of the IS style as evidenced by the two-tailed Sig values. As mentioned previously, given the rigor of training in the academic setting, this finding is also somewhat intuitive.

7.3: Data Analysis—TL Constituents

At this point, we will shift to discuss the approach taken to test the remaining research hypothesis (H0b), which asserts the following: There is no predominant leadership constituent style among actors with and without engineering degrees. Unlike the first hypothesis, here we are not looking to determine the extent to which the leadership styles differ when comparing the two groups (with and without engineering degrees). Instead the aim is to identify whether or not a predominant constituent style emerges within each group. In order to attempt this, three parametric statistical approaches were taken. First, each of the five constituent items for the TL leadership style previously discussed was compared against one another to determine if a difference existed in the mean perceptions for each group. Second, each of the perceived means for each overall leadership style (TL, XL, and PA) was compared with one another for differences in mean perceptions of the respective styles. Lastly, each of the leadership style scores for TL, XL, and PA were compared to the MLQ "norm" tables, which will be discussed later, and the extent that the scored style was (or was not) different from a given norm percentile was determined. This three-pronged approach is appropriate because, unlike the first hypothesis, the defined research focus here is on the full range of leadership to include XL and PA.

Regarding the five constituent test, restated we have the following: Ho: $\mu TL1 = \mu TL2 = \mu TL3 = \mu TL2 = \mu TL5$. The population means for all TL constituents with and without engineering degrees are the same. And conversely, we also have the following: Ha: $\mu TLcons_i \neq \mu TLcons_j$ for some $i \neq j$. The population means for at least two TL constituents with and without engineering degrees are different. ANOVA was employed to assess the first portion of this hypothesis regarding TL. And to allow for the maximum power, all TL constituent tests were conducted using the integrated data. As was the case with the t-test, the first step in the ANOVA analysis included review of a box plot to identify any outliers for the integrated sample population with engineering degrees. And as indicated in Figure 7.3-1, there were some outliers present. However, as mentioned previously, Elliot and Woodward had a similar view relative to ANOVA as with the t-test to which the author defers. More specifically, Elliot and Woodward (2007) cited Glass, Peckham, and Sanders and stated that "studies have shown the one-way ANOVA to be robust against some departures from assumptions ... if the sample size is large (at least 40) then the one sample t-test [or ANOVA] can be used without regard to skewness or outliers." Additionally, it was stated that "generally, non-normality of the data is not a concern unless you have small sample sizes or your data are highly non-normal ... if you have equal or near equal sample sizes, in each group, the equal variance assumption becomes less important." As mentioned previously, the author defers to these comments and proceeds with statistical testing employing ANOVA.

Figure 7.3-1: Five TL Constituent Means

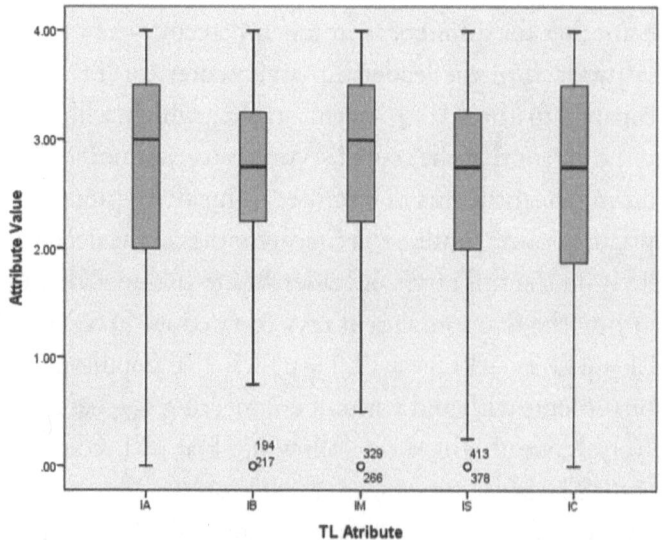

Referring to table 7.3-2, in terms of normality, none of the five constituent distributions met this criterion. However, as also indicated, the case size equaled 112. You should also note that the Sig values are 0.000. (This value is actually truncated at 0.000 but is equivalent to 0.0005.)

Table 7.3-2: Tests for Normality—TL Constituents

Tests of Normality

	TL Atribute	Kolmogorov-Smirnov[a]			Shapiro-Wilk		
		Statistic	df	Sig.	Statistic	df	Sig.
Attribute Value	IA	.137	112	.000	.931	112	.000
	IB	.135	112	.000	.938	112	.000
	IM	.146	112	.000	.916	112	.000
	IS	.112	112	.001	.950	112	.000
	IC	.153	112	.000	.946	112	.000

a. Lilliefors Significance Correction

Referring to table 7.3-3, the numerically highest perceived mean score is in the area of inspirational motivation (IM) with a mean of 2.82 and a confidence interval of 2.64 to 2.99.

Table 7.3-3: TL Constituent Descriptive Statistics

Descriptives

Attribute Value	N	Mean	Std. Deviation	Std. Error	95% Confidence Interval for Mean		Minimum	Maximum
					Lower Bound	Upper Bound		
IA	112	2.7768	.95822	.09054	2.5974	2.9562	.00	4.00
IB	112	2.7478	.88563	.08368	2.5819	2.9136	.00	4.00
IM	112	2.8192	.95520	.09026	2.6403	2.9980	.00	4.00
IS	112	2.6451	.93279	.08814	2.4704	2.8197	.00	4.00
IC	112	2.6161	.97371	.09201	2.4338	2.7984	.00	4.00
Total	560	2.7210	.94146	.03978	2.6428	2.7991	.00	4.00

Referring to table 7.3-4, you see that the test of homogeneity of variances yields a Sig value of 0.430. Because p = .430, which is greater than p = 0.05, Levene's test is not statistically significant, and the assumption of homogeneity of variances is therefore not violated. Referring now to the ANOVA result (shown in table 7.3-5), if the ANOVA is statistically significant (meaning that p less than .05), it can be concluded that not all group means are equal in the population (i.e., at least one group mean is different from another group mean). Alternatively, if p is greater than .05, no statistically significant differences exist between the group means. From the ANOVA, although numerical differences in means are evident, it may be concluded that there is no statistically significant differences between the group means at F(4, 555) = 0.955 and p = 0.432.

Table 7.3-4: Homogeneity of Variances Test

Test of Homogeneity of Variances
Attribute Value

Levene Statistic	df1	df2	Sig.
.959	4	555	.430

Table 7.3-5: Constituent ANOVA Results

ANOVA
Attribute Value

	Sum of Squares	df	Mean Square	F	Sig.
Between Groups	3.387	4	.847	.955	.432
Within Groups	492.079	555	.887		
Total	495.466	559			

Figure 7.3-6 shows the box plot for the integrated population without engineering degrees. As evident here, there are no outliers.

Figure 7.3-6: Integrated Box Plot Non-engineers

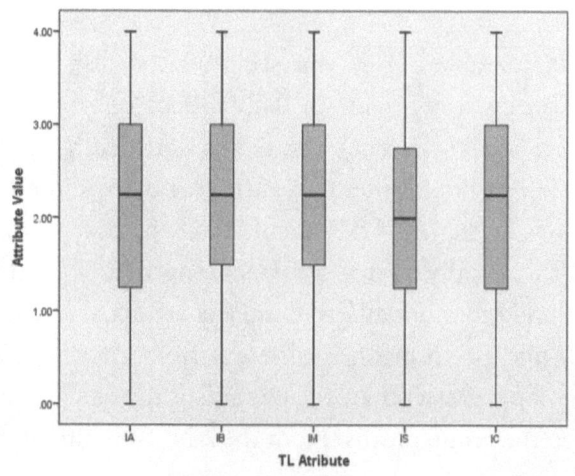

In terms of normality, as was the case with the integrated population with engineering degrees, those without engineering degrees are not normally distributed (see table 7.3-7). However, as will be shown in the descriptive statistics (see table 7.3-8), 169 cases comprised the sample set.

Table 7.3-7: Tests of Normality Non-engineers

Tests of Normality

	TL Atribute	Kolmogorov-Smirnov[a]			Shapiro-Wilk		
		Statistic	df	Sig.	Statistic	df	Sig.
Attribute Value	IA	.089	169	.002	.948	169	.000
	IB	.097	169	.001	.959	169	.000
	IM	.097	169	.001	.956	169	.000
	IS	.078	169	.013	.972	169	.002
	IC	.111	169	.000	.962	169	.000

a. Lilliefors Significance Correction

Table 7.3-8: Descriptive Statistics for Non-engineers

Descriptives

Attribute Value

	N	Mean	Std. Deviation	Std. Error	95% Confidence Interval for Mean		Minimum	Maximum
					Lower Bound	Upper Bound		
IA	169	2.1331	1.19212	.09170	1.9521	2.3142	.00	4.00
IB	169	2.1938	1.09270	.08405	2.0278	2.3597	.00	4.00
IM	169	2.2086	1.10251	.08481	2.0412	2.3760	.00	4.00
IS	169	2.0370	1.04285	.08022	1.8786	2.1953	.00	4.00
IC	169	2.0769	1.09823	.08448	1.9101	2.2437	.00	4.00
Total	845	2.1299	1.10608	.03805	2.0552	2.2046	.00	4.00

The numerically highest perceived mean score is, once again, in the area of inspirational motivation (IM) with a mean of 2.20 and a confidence interval of 2.04 to 2.37. Referring to table 7.3-9, it is evident that the test of homogeneity of variances yields a Sig value of 0.484. Consequently, Levene's test is not statistically significant, and the assumption of homogeneity of variances was not violated.

Table 7.3-9: Variances Test for Non-engineers

Test of Homogeneity of Variances

Attribute Value

Levene Statistic	df1	df2	Sig.
.867	4	840	.484

7.4: Data Analysis—Predominant Leadership Style

From the ANOVA (table 7.4-1), it may be concluded that there are no statistically significant differences between the group means with engineering degrees at $F(4, 840) = 0.749$ and $p = 0.559$.

Table 7.4-1: ANOVA for Non-engineers

ANOVA

Attribute Value

	Sum of Squares	df	Mean Square	F	Sig.
Between Groups	3.671	4	.918	.749	.559
Within Groups	1028.887	840	1.225		
Total	1032.558	844			

Refer once again to the predominant leadership style hypothesis (H0b) which asserts the following: There is no predominant style of leadership among actors with and without engineering degrees. For this second test, we are going to look at each of the three leadership styles (TL, XL, and PA) and seek to identify statistically significant differences within each group (those with engineering degrees and those without). This approach is reflected by the following restated hypotheses: Ho: $\mu TL = \mu XL = \mu PA$. The population means for each leadership style are equal. And, conversely, the approach is also reflected by the following: Ha: $\mu TL \neq \mu XL \neq \mu PA$. The population means for each leadership style are not equal. Figure 7.4-2 reflects the box plot for the population means.

Figure 7.4-2: Style Means Box Plot

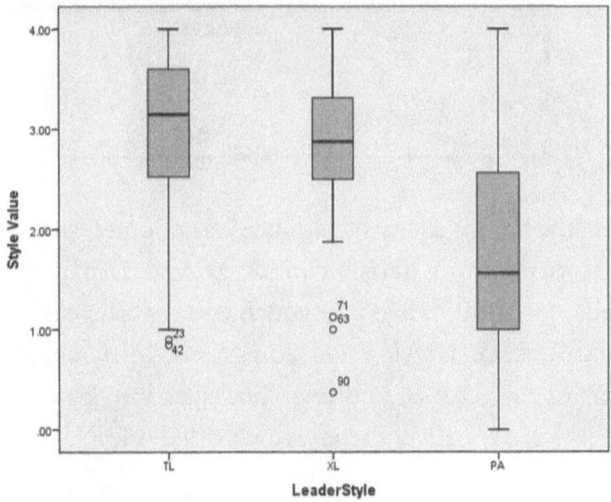

As indicated in table 7.4-3, because all P values in the "Wilk Sig" column are less than 0.05 except PA = 0.237, the style values are not all normally distributed.

Table 7.4-3: Normality Test for Leader Styles

Tests of Normality							
		Kolmogorov-Smirnov[a]			Shapiro-Wilk		
	LeaderStyle	Statistic	df	Sig.	Statistic	df	Sig.
Style Value	TL	0.13	48	0.042	0.899	48	0.001
	XL	0.127	48	0.052	0.937	48	0.013
	PA	0.091	48	.200[*]	0.969	48	0.237
*. This is a lower bound of the true significance.							
a. Lilliefors Significance Correction							

Referring to Table 7.4-4, which details the test of homogeneity of variances, because p = .042, which is less than 0.05, the Levene's test for homogeneity is statistically significant. Consequently, the assumption of homogeneity of variances was also violated.

Table 7.4-4: Leader-Style Variances Test

Test of Homogeneity of Variances			
Style Value			
Levene Sta	df1	df2	Sig.
3.248	2	141	0.042

Because the homogeneity of variance was not met, the output (table 7.4-5) must be used for decision making. And as p is less than .05 (actually equal to .0005), it can be concluded that there is a statistically significant difference in style value scores for the different levels of style applications (e.g., TL, XL, and PA). The question, however, is which leadership styles are different from which other ones?

Table 7.4-5: Robust Means Test of Leader-Style Means

Robust Tests of Equality of Means			
Style Value			
Statistic[a]	df1	df2	Sig.
Welch 24.321	2	92.695	0.0005
a. Asymptotically F distributed.			

SPSS offers post-hoc multiple comparisons to allow for the investigation of the differences pointed out in table 7.4-5. Referring now to table 7.4-6 (Post-Hoc Testing), as the Sig value or P value for TL compared to XL is greater than .05, the difference between these two group means is not statistically significant. As the Sig value for TL compared to PA is less than .05, the difference between these two group means is statistically significant. And as the Sig value (P value) for XL compared to PA is less than .05 (rather it is p = .0005), the difference between these two group means is also statistically significant.

Table 7.4-6: Post-Hoc Testing

Multiple Comparisons
Dependent Style Value   H2: CPMS WITH

(I) LeaderStyle			Mean Difference (I-J)	Std. Error	Sig.	Interval Lower Bound	Upper Bound
Tukey HSD	TL	XL	.13698	.18102	.730	-.2918	.5658
		PA	1.26198*	.18102	.000	.8332	1.6908
	XL	TL	-.13698	.18102	.730	-.5658	.2918
		PA	1.12500*	.18102	.000	.6962	1.5538
	PA	TL	-1.26198*	.18102	.000	-1.6908	-.8332
		XL	-1.12500*	.18102	.000	-1.5538	-.6962
Games-Howell	TL	XL	.13698	.16259	.678	-.2502	.5242
		PA	1.26198*	.19165	.000	.8052	1.7188
	XL	TL	-.13698	.16259	.678	-.5242	.2502
		PA	1.12500*	.18744	.000	.6780	1.5720
	PA	TL	-1.26198*	.19165	.000	-1.7188	-.8052
		XL	-1.12500*	.18744	.000	-1.5720	-.6780

*. The mean difference is significant at the 0.05 level.

Having run the same battery of tests for each of the leadership styles and for each of the subject groups with and without engineering degrees, tables 7.4-7, 8, and 9 summarize the resulting findings for those as well as for the CPM groups with and without engineering degrees. As indicated in table 7.4-7, there was a statistically significant difference between TL and PA as well as XL and PA in the CPM group with engineering degrees, but statistically significant differences were not detected between TL and XL in either group. Note that in the CPM group without engineering degrees, shown on the right-hand side of table 7.4-7, the same pattern exists with a statistically significant difference occurring between TL and PA as well as XL and PA, but no difference was detected between TL and XL.

Table 7.4-7: Post-Hoc CPM Means Comparison

	CPMS With & W/O Engineering Degrees	
Leader Styles	WITH	WITHOUT
TL	2.96	2.7
XL	2.83	2.48
PA	1.81	1.47

Taking a look at the manager results (see table 7.4-8), the same pattern was evident whereby in both groups (with and without engineering degrees), statistically significant differences occurred between TL and PA as well as XL and PA, but no statistically significant difference was detected between TL and XL.

Table 7.4-8: Post-Hoc Manager Means Comparison

Leader Styles	MANAGERS With & W/O Engineering Degrees	
	WITH	WITHOUT
TL	2.54	2.07
XL	2.43	2.03
PA	1.6	1.6

Taking a look at the integrated results (Table 7.4-9), as was the case above, the same pattern was evident whereby in both groups (with and without engineering degrees) statistically significant differences occurred between TL and PA as well as XL and PA, but no statistically significant difference was detected between TL and XL.

Table 7.4-9: Post-Hoc Integrated Means Comparison

Leader Styles	INTEGRATED With & W/O Engineering Degrees	
	WITH	WITHOUT
TL	2.72	2.13
XL	2.6	2.07
PA	1.64	1.59

To summarize this second test of the H0b hypothesis regarding a predominant leadership style, based on the foregoing analysis, the restated hypothesis must be rejected because of statistically significant differences in PA versus TL and XL among those with or without engineering degrees. However, as a practical matter, this is where a deeper understanding of the data becomes extremely important. From tables 7.4-7, 8 & 9, it is evident that the average values for PA are much lower than the same for TL or XL. This lower value is driven by the inherent design of the PA style construct. Indeed, when considering the 95th percentile norm table values for TL, XL and PA constituents (refer to Appendix D – table 1) it is clear that, in most cases, the listed values for TL and XL constituents are greater than those for PA constituents. Consequently, because the differences in TL and XL were not statistically significant, it would not be correct to assume that a predominant and unbiased leadership style is present among the tested actors.

7.5: Data Analysis—Percentile Comparisons

For the third test, referring once again to the predominant leadership style hypothesis, using the MLQ norm tables, the aim in this final test is to determine whether or not any perceived leadership style (TL, TX, or PA) is at a higher percentile level than any one of the remaining styles. This would be determined by comparing each of the population mean values to the "gold standard" value located in the norm tables (see appendix D). To explain how these tables are to be interpreted, the author refers to the specific MLQ score assessment recommendations offered by Mind Garden, the survey supplier. The first step is to group the like constituent items on the MLQ 5X (see appendix C), which simply sums their respective ratings and then divides them by the total number of items to get an average style constituent value. With this information in hand, Mind Garden suggests that the individual then be labeled more transformational or more transactional as opposed to simply stating that the individual be rated as either transformational or transactional. The averages for each style constituent and for the styles themselves are then compared to the norm tables. Referring to table

7.5-1 (Constituent Elements) as a point of clarification, recall that TL contains five constituent elements and that XL and PA contain only two items each respectively. And as previously mentioned, the items listed are to be summed and then divided by the total number of items to arrive at the constituent average. Accordingly, for each of the leadership styles (TL, XL, and PA), there are five, two, and two constituent elements respectively.

Table 7.5-1: Constituent Elements

Characteristic	Scale Name	Scale Abbrev	Items
Transformational	Idealized Attributes or Idealized Influence (Attributes)	IA or II(A)	10,18,21,25
Transformational	Idealized Behaviors or Idealized Influence (Behaviors)	IB or II(B)	6,14,23,34
Transformational	Inspirational Motivation	IM	9,13,26,36
Transformational	Intellectual Stimulation	IS	2,8,30,32
Transformational	Individual Consideration	IC	15,19,29,31
Transactional	Contingent Reward	CR	1,11,16,35
Transactional	Mgmt by Exception (Active)	MBEA	4,22,24,27
Passive Avoidant	Mgmt by Exception (Passive)	MBEP	3,12,17,20
Passive Avoidant	Laissez-Faire	LF	5,7,28,33

Table 7.5-2 is a reproduction of the norm table reflecting percentiles for subordinate ratings of higher levels with N equal to just more than twelve thousand cases. To ensure understanding, if a leader has a perceived IIA (idealized influence attributes) average rating of 3.00, he or she is operating in the fiftieth percentile. Likewise, if each of the five constituent elements of TL are averaged together (pooled), the overall perceived leader score can be determined on a percentile level. For example, if the fiftieth percentile scores for TL are all averaged, the mean TL score would then be 2.90. This 2.90 could be referred to as the gold standard for the fiftieth percentile TL rating based on the norming table.

Table 7.5-2: Norm Table

Percentiles for Individual Scores Based on Lower Level Ratings (US)

%tile	II(A) 12,118	II(B) 12,118	IM 12,118	IS 12,118	IC 12,118	CR 12,118	MBEA 12,118	MBEP 12,118	LF 12,118	EE 12,118	EFF 12,118	SAT 12,118	%tile
					MLQ Scores						Outcomes		
5	1.25	1.25	1.5	1.5	1	1.29	0.25	0	0	1	1.5	1	5
10	1.75	1.75	2	1.75	1.5	1.75	0.5	0	0	1.33	2	2	10
20	2.25	2.21	2.25	2.25	2	2.25	0.75	0.25	0	2	2	2.5	20
30	2.5	2.5	2.75	2.5	2.5	2.5	1.11	0.5	0.25	2.33	2.5	3	30
40	2.75	2.54	3	2.75	2.75	2.75	1.37	0.75	0.25	2.67	2.75	3	40
50	3	2.75	3	2.75	3	3	1.62	1	0.5	3	3	3.5	50
60	3.25	3	3.25	3	3.17	3.13	1.87	1	0.75	3	3.25	3.5	60
70	3.5	3.25	3.5	3.25	3.25	3.25	2.25	1.25	0.93	3.33	3.5	3.67	70
80	3.75	3.46	3.75	3.5	3.5	3.5	2.5	1.7	1.25	3.67	3.52	4	80
90	4	3.75	4	3.75	3.75	3.75	3	2	1.75	4	4	4	90
95	4	3.75	4	4	4	4	3.25	2.5	2	4	4	4	95

Table 7.5-3 reflects the survey reported mean scores for each of the three leadership styles from the perspective of subordinates as well as the respective percentiles for each of the row scores.

Table 7.5-3: Survey Reported Mean Scores

%tile	TL Mean	XL Mean	Pass Mean
5	1.3	0.77	0
10	1.75	1.13	0
20	2.19	1.5	0.13
30	2.55	1.81	0.38
40	2.76	2.06	0.5
50	2.9	2.31	0.75
60	3.13	2.5	0.88
70	3.35	2.75	1.09
80	3.59	3	1.48
90	3.85	3.38	1.88
95	3.95	3.63	2.25

Table 7.5-4 indicates the average scores from the received survey for the perceived leadership styles (TL, XL, and PA) of CPMs with (CPM W) and without (CPM WO) engineering degrees.

Table 7.5-4: CPM Percentile Levels

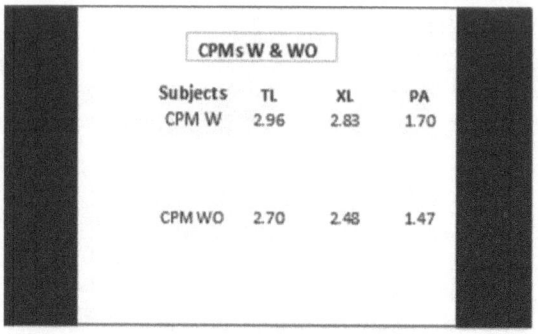

Based on the reported averages, the closest matching average leadership scores in the Mind Garden supplied norm tables (that were numerically less than the received scores) were then identified for each of the style scores as shown in table 7.5-3. As an example, the average perceived TL score for the CPMs with an engineering degree was 2.96, and the closest matching mean score from the norm table was found in the fiftieth percentile to be 2.90. However, although the average PA score for this same group was only 1.70, the closest corresponding norm table percentile that was also less than the received survey PA score was 1.48, which was located at the eightieth percentile.

The process then was to statistically compare the survey-reported average leadership style score to the closest not to exceed "Norm Table" match at an alpha of .05. If the survey-reported score was not statistically different from the mean score on the norm table, then the reported operating percentile level may also be assumed. However, if the survey-reported score was statistically different and numerically greater than the norm score, the operating percentile may very well have been higher than the percentile on the norm table. The one sample t-test was employed to implement the necessary comparisons.

The first test was for the CPM groups and TL. For simplicity of presentation, the box plots will not be shown. As indicated in table 7.5-5 (CPM Normality Test), based on the Wilk Sig test, given that the P value equals 0.001, which is less than 0.05, the reported TL data for CPMs with engineering degrees were not normally distributed.

Table 7.5-5: CPM Normality Test

Tests of Normality

	Kolmogorov-Smirnov[a]			Shapiro-Wilk		
	Statistic	df	Sig.	Statistic	df	Sig.
LdrStyle	.130	48	.042	.899	48	.001

a. Lilliefors Significance Correction

As reflected in table 7.5-6, the mean TL score (2.96 ± 0.82) was numerically higher than the population fiftieth percentile TL score of 2.90 as demonstrated previously.

Table 7.5-6: TL Mean Score

One-Sample Statistics

	N	Mean	Std. Deviation	Std. Error Mean
LdrStyle	48	2.9625	.82019	.11838

However, referring to table 7.5-7 (One Sample t-Test Results), the TL score was not statistically significantly different from the population fiftieth percentile score, t(47) = .528, and p = .600. Because the reported score was not statistically different from the "gold standard" percentile score, there was no statistical basis for rejecting the theory that the perceived demonstration of TL was not equal to the fiftieth percentile level from the norm table.

Table 7.5-7: One Sample of t-Test Results

One-Sample Test

	Test Value = 2.90					
	t	df	Sig. (2-tailed)	Mean Difference	95% Confidence Interval of the Difference	
					Lower	Upper
LdrStyle	.528	47	.600	.06250	-.1757	.3007

The results discussed in tables 7.5-5, 7.5-6, and 7.5-7 as well as the results for the remaining tests for the two CPM groups (with and without engineering degrees) are shown in table 7.5-8. In all cases, application of the one sample t-test did not identify statistically significant differences between the percentile from the norm table and the perceived operating level of the CPMS with or without engineering degrees. Consequently, in both groups, the highest operating percentile scores were related to the PA leadership style.

Table 7.5-8: CPM Percentile Results

	Subjects	TL	XL	PA
	CPM W	2.96	2.83	1.70
%Tile	50	2.90		
%Tile	70		2.75	
%Tile	80			1.48
	CPM WO	2.70	2.48	1.47
%Tile	30	2.55		
%Tile	50		2.31	
%Tile	80			1.48

Referring to table 7.5-9, while no statistically significant differences were revealed for managers with engineering degrees versus the gold standard listed in the norm table for the TL, XL and PA styles, statistically significant differences were identified for managers without engineering degrees for TL and XL but not for PA. In both cases, the differences point to the managers operating above the selected norm percentile.

Table 7.5-9: Manager Percentile Results

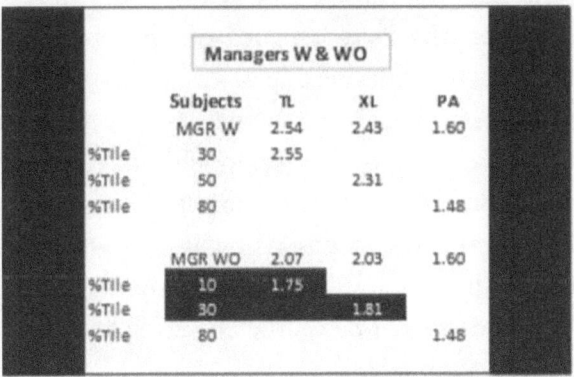

	Subjects	TL	XL	PA
	MGR W	2.54	2.43	1.60
%Tile	30	2.55		
%Tile	50		2.31	
%Tile	80			1.48
	MGR WO	2.07	2.03	1.60
%Tile	10	1.75		
%Tile	30		1.81	
%Tile	80			1.48

Referring to table 7.5-10, although no statistically significant differences were revealed for the integrated population compared to the gold standard on the norm table with engineering degrees in the XL and PA styles, a statistically significant difference was identified for this group in the TL style. Likewise, in the integrated population without engineering degrees, both the PA and XL leadership styles were not statistically significantly different from the norm table while the TL style was statistically different from the norm table in this group. In both cases, the TL differences point to the integrated population operating above the selected norm percentile.

Table 7.5-10: Integrated Percentile Results

	Subjects	TL	XL	PA
	INTGR W	2.72	2.60	1.64
%Tile	30	2.55		
%Tile	60		2.50	
%Tile	80			1.48
	INTGR WO	2.13	2.07	1.59
%Tile	10	1.75		
%Tile	40		2.06	
%Tile	80			1.48

Summarizing the findings presented in the preceding tables about CPMs, the one sample t-test did not identify statistically significant differences between the gold standard on the norm table and the reported perceived mean leadership style scores (with and without engineering degrees). Regarding managers, the one sample t-test identified differences only in this group without engineering degrees and for the TL and XL styles. Regarding the integrated population, the one sample t-test revealed differences in this group regarding TL with engineering degrees and TL without engineering degrees. The differences suggested higher percentile operating levels versus the gold standard from the norm table.

Refer to figure 7.5-11, which compared the integrated population (including CPMs and non-CPM managers) to the fiftieth percentile norm score. A key observation here, as was identified in the second test in the previous sections, is that the integrated population with engineering degrees appears to be operating at an overall higher percentile level than those without engineering degrees. There is also a slight yet obvious downward trend in the integrated population with and without engineering degrees from TL to XL and then more decelerated to PA. And although the trend might suggest both groups (with and without engineering degrees) operate at a higher level of TL when compared to the other leadership styles, there is still insufficient evidence to reject the hypothesis that there is no predominant leadership style among those with or without engineering degrees. This, of course, is due to the lack of statistical significance between TL and XL. Although there were statistically significant differences identified for the mean PA leadership style when compared to TL or XL, it is just not practical to conclude that a predominant style exists as the PA scores are also very different from TL and XL in the norm tables. In both comparisons, the mean PA scores are significantly and numerically lower than either TL or XL.

Figure 7.5-11: Reported Scores and the Fiftieth Percentile Norm Table

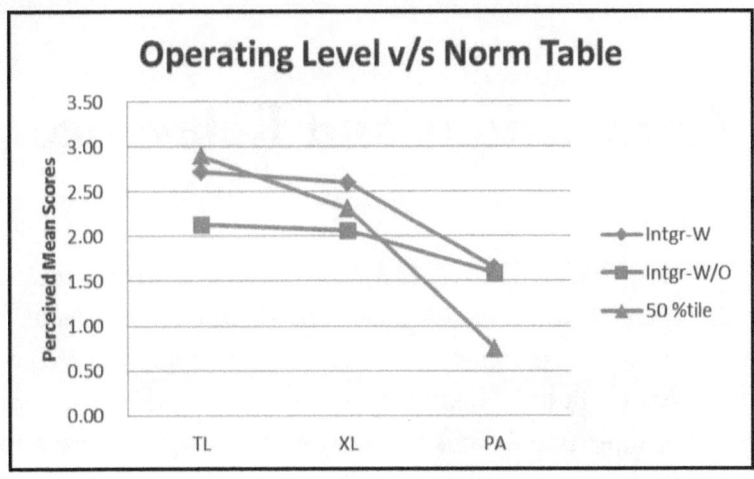

Summarizing all testing for the second hypothesis, for the first test, the ANOVA test did not identify statistically significant differences at the 95 percent confidence level between the five TL constituents (IA, IB, IM, IS, and IC). For the second test, the ANOVA identified statistically significant differences at a 95 percent confidence level between TL and PA as well as XL and PA. However, TL and XL were not statistically different. For the third and final test, the one sample t-test confirmed that for all groups with and without engineering degrees, at the 95 percent confidence level, varying %tile levels of "in-group" demonstration of full-range leadership styles (TL, XL, and PA) were perceived to be present.

CHAPTER 8

Conclusions and Relevance

THE FIRST RESEARCH hypothesis (H0a) states the following: There is no relationship between engineering education and transformational leadership (TL). Based on the evidence presented, this hypothesis should be rejected when considering the manager and integrated sample populations with engineering degrees versus those without. Of course, this suggests that those with engineering degrees are more transformational. However, perhaps because of the reduced power of the test, analysis of the CPM groups did not identify statistically significant differences at an alpha of 0.05.

Another statistically significant difference, this one occurring in the manager and integrated populations with engineering degrees, is a higher perceived level of XL style versus the same for those without engineering degrees. On the surface this difference may appear to undermine the significance of the TL findings for the same group. This should not be the case, however, when considering a couple of key mitigating factors. First, it is incumbent upon leaders to make clear the expectations (e.g., goals and objectives) for subordinates, which may also be viewed in the context of providing what to do. How and why subordinates achieve the goals and objectives may be linked to, among other things, motivation and inspiration provided by the leader. As also discussed in chapter 3, according to Avolio and Bass (2004), some of the qualities associated with XL include "provides assistance in exchange for efforts, discusses who is responsible for what, makes clear [the] rewards for efforts, focuses attention on mistakes and attention [is] directed to failure." TL qualities include: "inspire, instill pride, sense of purpose, displays confidence, talks optimistically, articulates a vision [and] questions assumptions."

From these comments it should be clear that effective leaders must provide both what is to be done and concurrently offer vision and strategies regarding how such may be accomplished. Supporting this point, Avolio and Bass (2004) stated that "the transactional process, [contingent reward] in which the leader clarifies what the associates need to do for a reward, is nevertheless viewed ... as an essential component of ... effective leadership." Second, contingent reward is one of only two XL constituents, thereby accounting for 50 percent of the total perceived style rating. The other constituent for XL is management-by-exception: active (MBEA). Bennet (2009) cited works of multiple authors who argued that contingent reward was in itself related to TL. Thus, it might be concluded that if the reported perceived contingent reward (CR) constituent of XL is, in essence, driving the overall XL mean score and XL is determined to be statistically different and higher for those in the integrated population versus the same without, such may be consistent with previous arguments posed by Avolio, Bass, and Bennet, namely that CR in combination with TL may be required for effective leadership. Table 8-1 reflects the mean scores for CR and MBEA.

Table 8-1: Perceived Mean Scores for CR and MBEA

Group Statistics

	LDRSTYLE	N	Mean	Std. Deviation	Std. Error Mean
SCORE	INTXLCR	112	2.8058	0.8461	0.07995
	INTXLMBEA	112	2.3884	0.93472	0.08832

To test this theory, an independent sample's t-test was implemented to identify mean differences for both XL constituents (CR and MBEA shown in table 8-2). Referring to table 8-2, it can be seen that the normality assumption was not met for either group. The P values of .001 and 0.015 for CR and MBEA respectively were less than .05.

Table 8-2: CR and MBEA Means Test

		Kolmogorov-Smirnov[a]			Shapiro-Wilk		
	LDRSTYLE	Statistic	df	Sig.	Statistic	df	Sig.
SCORE	INTXLCR	0.091	112	0.023	0.952	112	0.001
	INTXLMBE	0.081	112	0.069	0.971	112	0.015

a. Lilliefors Significance Correction

Referring now to Table 8-3, because the Sig value equals .280 and this is > p = .05, the variances are equal. Also note the value in the sig (two-tailed) column in the "Equal Variances Assumed" row. Because p = .001 is less than .05, it can therefore be concluded that CR and MBEA do have statistically significantly different mean XL constituent style scores with the CR mean being numerically greater. This difference in CR and MBEA scores may potentially support previously referenced arguments suggesting that the CR component of XL is linked to TL and consequently, effective leadership.

Table 8-3: Equal Variance Test for CR and MBEA

Independent Samples Test

		Equality of Variances		t-test for Equality of Means					Interval of the	
		F	Sig.	t	df	Sig. (2-tailed)	Mean Difference	Std. Error Difference	Lower	Upper
SCORE	Equal variances assumed	1.278	.260	3.504	222	.001	.41741	.11913	.18263	.65219
	Equal variances not assumed			3.504	219.833	.001	.41741	.11913	.18262	.65220

Given the results identified in researching the first hypothesis, it begs the question why engineers shy away from human issues. To this point, Bonasso (2001) suggests that "most people who know engineers would say it's because social issues tend to be gray and vague and not objective, verifiable, and defensible ... engineers have no formulas or numbers with which to conceptualize these issues, so they do not enter the fray."

The second research hypothesis (H0b) asserted the following: There is no predominant style of leadership among actors with and without *engineering degrees.* Efforts to assess this hypothesis required a three-pronged statistical approach, including ANOVA and the one sample t-test. Based on the evidence presented from the first test and at an alpha level of 0.05, no statistically significant constituent differences were detected for TL for the integrated population. Likewise, as determined by the second test, TL and XL were not statistically different when comparing the reported mean leadership styles for all three sample populations (CPMs, managers, and integrated). Finally, although a visible trend existed in the integrated population (for those with and without engineering degrees) from the TL style downward to the PA style, there was no statistically significant difference at an alpha level of 0.05 for the TL and XL perceived mean scores. Resulting from the lack of statistical significance here, and considering the previous practicality discussion regarding PA and its respective inherently low mean scores relative to the remaining style mean values, rejection of this hypothesis should not suggest the identification of a predominant leadership style. I'd like to offer a note to the reader at this point regarding the second hypothesis.

With so much statistical information provided, it might be very easy to get a bit wrapped around the axle when contemplating the difference between the two hypothesis tests. The test currently being discussed (the second test) sought to identify a predominant leadership style (TL, XL, and PA) for those populations with or without engineering degrees. To this point, if you are a degreed engineer, is your leadership style more transformational, transactional, or passive avoidant? Likewise, if you are not a degreed engineer, is your leadership style more transformational, transactional, or passive avoidant? Recall that the first hypothesis sought to determine whether degreed engineers' leadership styles differed from the leadership styles of those without engineering degrees. The relevance of this approach is supported by the notion that simply knowing whether or not an engineer with a degree or a leader without one is predominantly operating at the TL level when compared to other leadership styles (XL or PA), reveals nothing regarding which category (degreed or non-degreed) is more or less transformational, transactional, or passive avoidant.

Further buttressing research results, the reader may recall the author's deference to comments and citations offered by Elliot and Woodward (2007) regarding the severity of assumptions (e.g., normality, outliers, etc.), namely that the parametric tests employed were robust enough to accommodate some departures from these assumptions while still providing valid statistical results. In an effort to offer further support for this position, two nonparametric tests were run—the Mann-Whitney U and the Kruskal-Wallis. The "Mann-Whitney U (compare[s] two independent groups) [and served as a] nonparametric alternative to a two sample t-test" (Elliot and Woodward, 2007). Likewise, the "Kruskal-Wallis (compare[s] two or more independent groups) [and served as a] nonparametric alternative to a one-way analysis of variance" (Elliot and Woodward, 2007). The reader may refer to appendix G to view these applications.

8.1: Generalizability

There are limitations with regard to the generalizability of the preceding research, which aimed to determine the relationship—if one existed—between engineering education and leadership style with an emphasis on TL. And while some statistically significant differences were detected, particularly in the larger populations, such should not be interpreted to suggest generalizability to all those with engineering degrees. Indeed, literature abounds regarding the lack of leadership skills, perhaps because of the lack of desire for such positions among engineering graduates. What can be said of the generalizability of the results is that predicated on the sampled integrated population, inclusive of those with and without engineering degrees who held leadership positions, raters perceived the group with engineering degrees to be more transformational and transactional than those without.

The design of the MLQ5X is also helpful in providing some insight as to the overall leadership effectiveness resulting from the perceived mean style scores. However, although such data were also collected with the current research, addressing this area was not within the research scope.

8.2: Limitations and Areas for Future Research

The current research also identified opportunities for future research. Returning to some of the demographic information reviewed earlier, although any comprehensive analysis was well beyond the scope of the current research, ANOVA and the independent samples t-tests were implemented considering the integrated population (with and without engineering degrees) to determine the extent to which experience, gender, and the number of organizational reports (direct and/or matrix) may have influenced the perceived leadership style of the individual being rated. As is evident from the results in table 8.2-1, females were different from males (female mean 2.06 versus male mean 1.57). Otherwise, no statistically significant differences were identified between the various demographic and organizational structure factors for either of the integrated population groups (with or without engineering degrees).

Table 8.2-1: Areas for Future Research

Integrated with engineering degrees	The TL Value was NOT statistically significantly different between experience Levels, $F(2, 109) = 2.538$, $p = 0.084$.
	The TL Value was NOT statistically significantly different between Gender Levels, $T(109) = -.543$, and $p = .588$.
	TL Value was NOT statistically significantly different between Direct Report Levels, $F(2, 109) = 2.2268$, $p = 0.113$.
Integrated without engineering degrees	The TL Value was NOT statistically significantly different between Experience Levels, $F(2, 166) = 3.401$, $p = 0.036$.
	The TL Value **WAS** statistically significantly different between Gender Levels, $T(137) = 3.119$, and $p = .002$.
	The TL Value was NOT statistically significantly different between Direct Report Levels, $F(2, 172) = 2.526$, $p = 0.083$.

Based on the previously outlined preliminary results, one may pose the following future research questions:

1. Is the perception of leadership style of those with engineering degrees influenced by leader gender or experience?
2. Is the perception of leadership style of those with engineering degrees influenced by the number of reports?
3. Does the possession of an engineering degree and/or experience by the rater influence the perception of leadership style?
4. Does the possession of an engineering degree by the rater and/or the rater's gender influence the perception of leadership style?

As mentioned in the research limitations section, another future research opportunity is to determine the perceived leadership effectiveness based on data collected while also considering the questions previously posed. With answers to the expanded questions and leadership effectiveness, the generalizability of the current research may be further substantiated.

8.3: Where Do We Go from Here?

Based on the research results, we know that those serving in leadership positions who also possess engineering degrees are perceived to be both more transformational and transactional than those who do not possess engineering degrees. In the next chapter, we will explore how this finding may have facilitated the career ascension of Josef Brilliant, an electrical engineer employed by the hypothetical Summit Consumables Corporation.

CHAPTER 9

Meet Josef Brilliant

PRIOR TO DISCUSSING Joseph Brilliant, I'd like to briefly introduce you to Summit Consumables Incorporated, which has been in the business of making and selling consumer products for decades. And like most large corporations, Summit Consumables has shifted its growth strategy many times, trying both organic and inorganic approaches. Over the last few years, Summit Consumables has been receiving pressure from its competition, especially from a relatively new and more efficient and agile competitor. The stock price has begun to take a hit, and based on revenue declines, the company is contemplating reducing its current dividend yield. Analysts are critiquing Summit Consumables's cost structure and are suggesting that the company could do a great deal more to get costs in line with decreasing revenues. Moreover, compared to the industry average, Summit Consumables's P/E ratio of 42.5 is excessive. A snapshot of the corporate financial structure is reflected in figure 9-1

Figure 9-1: Summit Consumables Financial Snapshot

The Corporation: Consumer Products	
Market Capitalization	$105 billion
Shares Outstanding	$ 1.4 billion
Current Stock Price	$ 75.09
Dividend Yield	0.035
P/E (Trailing 12 months):	42.5x

Some of the changes discussed in the boardroom following the last earnings report (which missed analyst estimates yet again) include the following:

- slashing the workforce by 10 percent,
- modifying the current executive compensation program,
- eliminating cash bonuses for employees reporting more than four levels from the regional president,
- expanding the managerial span of control to a minimum of ten direct reports,
- reducing operations costs by 10 percent, and/or
- delaying all non-vital capital projects until the year's end.

Despite the tough talk about performance and competition, the message from headquarters is that the company vows to fully reward employees who demonstrate leadership in all areas of their responsibility. Just before kickoff of the annual performance review cycle, Chairman Dave Sterling of Summit Consumables sent out an email to all salaried employees containing an article regarding transformational leadership that included the following excerpt and comment:

> [Avolio and Bass reported that] when all levels of managers, students, and project leaders around the world were asked to describe the characteristics and behaviors of the most effective leaders with whom they had worked in the past, the characterizations were more transformational than transactional. Among the specific descriptors used for these leaders were "inspirational, intellectually stimulating, challenging, visionary, development oriented, and determined to maximize performance" (2004). We believe that these characterizations essentially mirror the five constituent elements of TL. Thus, the presence of these attributes in any one or more of the tested groups might also identify a leadership feeder pool for future effective leaders.

Headquartered in Wilmington, Delaware, Summit Consumables also has physical manufacturing and distribution locations in Texas, Georgia, Kentucky, and New Jersey. Leading the operations of each of these facilities is a regional president. Services such as product-quality monitoring, environmental engineering, employee safety, legal support, human resources, and community relations are sourced from headquarters. There has been a great deal of buzz regarding the potential closure of one of the facilities located in the South, resulting in operations consolidation and potential incremental job loss beyond the 10 percent that was previously contemplated. Summit Consumables continues to use the annual performance appraisal approach, and the word from the executive level is that "based on the current performance appraisal, all employees need to fully understand the gravity of the competitive landscape."

Jessica Wright (manufacturing director), Johnny Goode (senior buying analyst), Kimberly Hours (regional sales director), Maria Summers (regional product portfolio director), Earl Easy (IT manager), Kenneth Bethone (call center manager), and Josef Brilliant (senior electrical engineer) are all employed at the Georgia facility. All have just received their respective performance evaluations, and the pressure is on to turn the facility around in hopes that it will withstand the external competitive environment. Summit Consumables uses a five-point rating system, with a five representing the highest available performance rating and a one the lowest. As a note to the reader, I discuss in detail the financial concepts presented in this section and others (e.g., P/E ratio, market capitalization, NPV, IRR, budgeting, etc.) as well as provide a detailed analysis of the performance appraisal process and each of the other Summit Consumables employees in my book titled *All the Way to The Top: A Practical Guide to Corporate and Business Leadership*. While I will not address them in detail here, I do encourage you to read that title.

Let's spend a bit of time reviewing Josef's job assignments and performance results. Josef Brilliant services the entire facility. Although he does not have any solid line (direct) reports, he does serve as the team leader (on a matrix basis) for all maintenance and engineering projects. One such project was the recent modernization effort to replace all equipment with newer and faster machines that were capable of responding to the increased changeovers required to accommodate R & D's recent influx of new product trials. Figure 9-2 reflects his location in the organization. This is his fourth year with Summit Consumables after joining the company as an entry level engineer. Having a bachelor of science in electrical engineering and a minor in engineering management, Josef has openly expressed his desire to progress into management, preferably in the production area.

Figure 9-2: Josef's Organization Structure

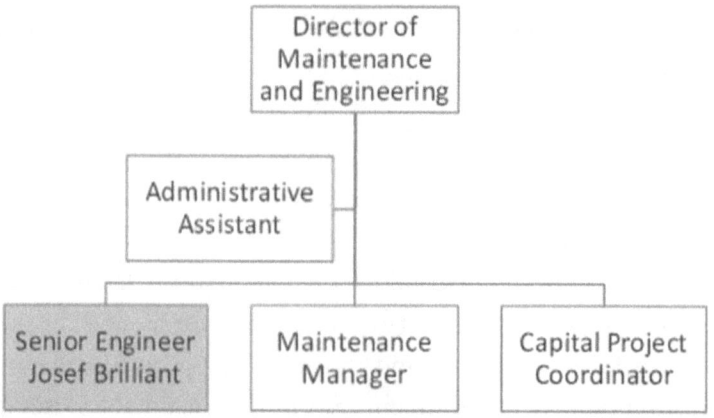

Josef has received a four rating in two of his last three ratings, and the last of these ratings was a three. Let's now take a look at his current feedback and performance rating.

As indicated in Figure 9-3, Joseph's performance rating was a three for the current period. This rating was also below his expectations as he had been told by a prior manager that he would need to get at least two consecutive four ratings out of the last three evaluations in order to be considered for another position (either promotional or developmental). Josef's current director, Daniela, thinks that he has a great deal of potential but that he needs to, in her words, "ease up" and not appear to be "on edge" all the time. The new equipment installation has begun to slip on the schedule, and there has been some talk of having an engineer from the central group in headquarters "look over Josef's shoulder" to get things back on track. She further indicated that despite his excellent problem-solving skills, her personal observations of his communication skills, which were also supported by feedback from multiple unsolicited sources, suggests that clarity and brevity would be extremely helpful in getting him a better rating going into future performance periods. Daniela ended the meeting with Josef stating that "even the most brilliant engineers can run into significant career head winds if they are unable to articulate to the lay community."

Figure 9-3: Excerpt from Joseph's Performance Evaluation

Performance Evaluation: Josef Brilliant
Overall Rating 3 - Satifactory

Key Strengths: Intelligence, commitment, customer focus

Development Opportunities: Communications (written and oral), does not receive feedback well.

Supervisor recommendations: Consider being more in touch with your emotions and how your actions might detract from your work.

This was not a bad performance review after all, and Josef is viewed as a satisfactory contributor. However, Josef is not content with being satisfactory. He aspires to progress up the corporate ladder. And considering Josef's feedback from the performance appraisal, there are several strengths that he may leverage to actually build upon his development opportunities. Before we see how Josef responds to the feedback, let's first address a topic that has frequently plagued engineering professionals—communications. The notion that communications is an essential function of an engineer's work environment is not at all new. As pointed out by Romanowski and Sageev (2001), referring to results of a May 1999 survey, "The overwhelming evidence shows that graduating engineers are inadequately equipped to meet this [communication] need ... numerous industry surveys, managers' comments, and academic studies confirm this assessment." Hartman and Jahren (2015) conducted research with the aim (purpose) as follows: "The purpose of this qualitative research study is to identify the most important leadership competencies undergraduates seeking full-time employment should possess when applying for positions." One key finding from that research was that "all interviewees discussed the importance of communication skills. Included in this theme were written, oral, non-verbal, and listening skills, as well as the ability to conduct crucial conversations." More contemporary, according to Adams and Keshwani (2017), "Engineers report that communication is an important activity that consumes approximately half of their day requiring conversation with both technical and non-technical audiences." According to the National Academy of Engineering, as cited by Brazil and Farr (2009), "Technical excellence is the essential attribute of engineering graduates, but those graduates should also possess team communication ... skills." Literature also abounds regarding efforts to build development of communication skills into engineering curricula. At the same time, there is perhaps no one more qualified to speak about the technical aspects of how and why things work the way they do than engineers. As we saw earlier in this book, according to Yao and Russel (1997), "An engineer is hired for her or his technical skills, fired for poor people skills, and promoted for leadership and management skills."

9.1: Communication—It's More Than Talking

The following discussion is based largely on a segment from material that I use in the delivery of an effective communications seminar. I have tailored it slightly for our purposes here. A very large part of what we do when we are not asleep is communicating with others. Whether it's sharing a project update with your boss, the level of attention that you give to your spouse when he or she is talking to you, or the look that you give to a passing motorist, it all means something. *Merriam-Webster (2007)* defines communication as "the act or process of using words, sounds, signs, or behaviors to express or exchange information or to express your ideas, thoughts, feelings, etc., to someone else. : a message that is given to someone : a letter, telephone call, etc." Executing communication typically occurs in one of the following three channels:

1. **Spoken:** There are two components to spoken communication. There's verbal communication. This is what you are saying. You have no doubt heard someone reference a *verbal agreement*. And there's paraverbal. This refers to how you say something—your tone, speed, pitch, and volume.
2. **Nonverbal:** These are the gestures and body language that accompany your words. Some examples include your arms folded across your chest, tracing circles in the air, tapping your feet, or having a hunched-over posture.
3. **Written:** As previously pointed out, communication can also take place via fax, email, or otherwise in writing.

On the surface, communication seems pretty simple. I talk, and you listen. You send me an email, and I read it. However, effective communication is far more complicated than it seems. This is especially true when highly technical people (e.g., engineers) attempt to communicate with an audience that has little to no training or experience in the technical area. In some respects, the engineer may be perceived as speaking a different language altogether. You may think of this as a sort of cultural difference. Thus, engineers should always know their audience and gauge how much technical jargon if any is appropriate when engaging with their audience.

Consider the saying "It's not what you say, but it's how you say it." It's true! Try saying these three sentences out loud, placing the emphasis on the italicized word.

- "*I* didn't say you were wrong." (This implies that it wasn't me.)
- "I didn't *say* you were wrong." (This implies that I communicated it in another way.)
- "I didn't say you were *wrong*." (This implies that I said something else.)

Paraverbal communication consists of three basic parts of the message as reflected by the pitch, tone, and speed of chosen words when we communicate. Pitch can be most simply defined as the key of your voice. A high pitch is often interpreted as anxious or upset. A low pitch sounds more serious and authoritative. People will pick up on the pitch of your voice and react to it. So too, variation in the pitch of your voice is important to keep the other party interested. If you naturally speak in a very high-pitched or low-pitched voice, work on varying your pitch to encompass all ranges of your vocal cords. (One easy way to do this is to relax your throat when speaking.) Make sure to pay attention to your body when doing this. You don't want to damage your vocal cords.

Did your mother ever say to you, "I don't like that tone!" She was referring to the combination of various pitches to create a mood. (Speed, which we will see here, can also have an effect on your tone.) Here are some tips on creating a positive, authoritative tone.

- Try lowering the pitch of your voice a bit.
- Smile! This will warm up anyone's voice.
- Sit up straight and listen.
- Monitor your inner monologue. Negative thinking will seep into the tone of your voice.

The pace at which you speak also has a tremendous effect on your communication ability. From a practical perspective, someone who speaks quickly is harder to understand than someone who speaks at a moderate pace. Conversely, people who speak very slowly will probably lose their audience's interest before they get very far. Speed also has an effect on the tone and emotional quality of your message. A hurried pace can make the listener feel anxious and rushed. A slow pace can make the listener feel as though your message is not important. A moderate pace will seem natural, and it will help the listener focus on your message. One easy way to check your pitch, tone, and speed is to record yourself speaking. Think of how you would feel listening to your own voice. Work on speaking the way you would like to be spoken to.

Nonverbal communication is just that. When you are communicating, your body is sending a message that is as powerful as your words. In our following discussions, remember that our interpretations are just that—common interpretations. (For example, the person sitting with his or her legs crossed may simply be more comfortable that way and not feel closed-minded toward the discussion. Body language can also mean different things across different genders and cultures.) However, it is good to understand how various behaviors are often seen so that we can make sure our bodies are sending the same messages as our mouths.

Think about these scenarios for a moment. What nonverbal messages might you receive in each scenario? How might these nonverbal messages affect the verbal message?

- Your boss asks you to come into his or her office to discuss a new project. The boss looks stern, and his or her arms are crossed.
- Some team members tell you that they have bad news, but they are smiling as they say it.
- You tell a few coworkers that you cannot help them with a project. They say that it's okay, but they slam your office door on their way out.

In 1971, psychologist Albert Mehrabian published a famous study called "Silent Messages." In it, he made several conclusions about the way the spoken word is received. Although this study has been misquoted often throughout the years, its basic conclusion is that 7 percent of our message is verbal, 38 percent is paraverbal, and 55 percent is from body language. Now we know this is not true in all situations. If people are speaking to you in a foreign language, you cannot understand 93 percent of what they are saying. Or if you are reading a written letter, you are likely getting more than 7 percent of the sender's message.

What this study does tell us is that body language is a vital part of our communication with others. With this in mind, let's look at the messages that our bodies can send. Body language is a very broad term that simply refers to the way in which our body speaks to others. I have included an overview of three major categories here, and we will discuss a fourth category (gestures) in a moment.

1. *Consider the way that you are standing or sitting.* Think for a moment about different types of posture and the message that they relay. Sitting hunched over typically indicates stress or discomfort. Leaning back when standing or sitting indicates a casual and relaxed demeanor. Standing ramrod straight typically indicates stiffness and anxiety.
2. *Consider the position of your arms, legs, feet, and hands.* Crossed arms and legs often indicate a closed mind. Fidgeting is usually a sign of boredom or nervousness.
3. *Consider your facial expressions.* Smiles and frowns speak a million words. A raised eyebrow can mean inquisitiveness, curiosity, or disbelief. Chewing one's lips can indicate thinking, or it can be a sign of boredom, anxiety, or nervousness.

A gesture is a nonverbal message that is made with a specific part of the body. Gestures differ greatly from region to region and from culture to culture. Table 9.1-1 provides a brief list of gestures and their common interpretation in North America.

Table 9.1-1: Common Interpretation of Gestures

Gesture	Interpretation
Nodding head	Yes
Shaking head	No
Moving head from side to side	Maybe
Shrugging shoulders	Not sure; I don't know
Crossed arms	Defensive
Tapping hands or fingers	Bored, anxious, nervous
Shaking index finger	Angry
Thumbs up	Agreement, OK
Thumbs down	Disagreement, not OK
Pointing index finger at someone/something	Indicating, blaming
Pointing middle finger (vertically)	Vulgar expression
Handshake	Welcome, introduction
Flap of the hand	Doesn't matter, go ahead
Waving hand	Hello
Waving both hands over head	Help, attention
Crossed legs or ankles	Defensive
Tapping toes or feet	Bored, anxious, nervous

9.2: Effective and Efficient Communications

You have no doubt heard that you only have so much time to get your point across. Thus, your message has to be crisp, clear, and to the point. With this in mind, another focus area for the communications aspects of my leadership development seminars is how to get the attention of others while also hitting the mark with your conversation. As engineers, we find familiarity with structure and processes, so let's consider that context for this segment. Figure 9.2-1 depicts a structured communications process flow from left to right. In addressing each of the four stages, the key is to leverage the six roots of open questions (who, what, when, where, why, and how). Implementation of the structured flow requires us to complete the first stage prior to moving to the second stage and so on until the end.

Figure 9.2-1: Structured Communications Process

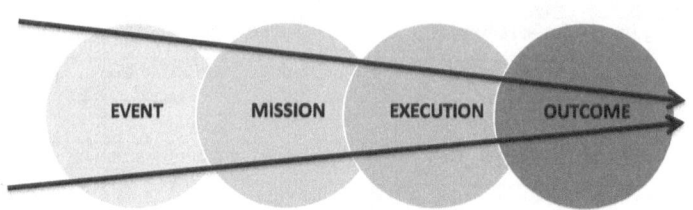

The objective of stage one is to clearly state the event. Try to make this no longer than one sentence. If you are having trouble, ask yourself, "Where? Who? When?" This will provide the initial framework for your message so it can be clear and concise. An example of communication in this stage could be as follows: "Yesterday, I was in process controls meeting at the flavors plant." Next, briefly state the mission—what you did or what took place and why if applicable. Again, this should be no longer than one sentence. Continuing with the previously outlined example, you might say, "I was asked to discuss my recommendation for resolving a processing problem." Now using the question *how* to frame your next few words, state what you did to resolve the problem or issue in one sentence. The goal of the execution part of the process is to provide a solid description and state the precise steps taken to resolve the issue. An example here might be as follows: "I used the whiteboard to sketch out my proposed solution." Keep in mind that the main reason you are sharing this information with someone else is to lead up to some end or outcome. With that in mind, you want to state the outcome in the end. This will often use a combination of the six roots mentioned previously. Again, a precise short description of the outcome from your previous steps will complete the final process stage. You might conclude by saying, "After fielding a few questions, the team bought into my proposal and asked me to flesh out the details."

Now let's bring all the process pieces together using a realistic scenario. You and a fellow engineering colleague have met for lunch on Monday. She matter-of-factly asks, "How is everything going? Anything exciting happen to you lately?" You then respond, "Yesterday I was in a process controls meeting at the flavors plant, and I was asked to discuss my recommendation for resolving a processing problem. I used the whiteboard to sketch out my proposed solution. The team bought into my proposal and asked me to flesh out the details."

This format can be compressed for quick conversations or expanded for lengthy presentations. The key point, however, is that you speak in a cogent and complete manner in order to ensure that you cover the key points that you desire to convey.

You should also be aware that efficient communication is just as important as effective communication. While the words effective and efficient are very close in meaning, there is a subtle difference as used here. Although both effective and efficient, when used with regard to communications (oral or written), address hitting the mark, efficient goes a step further as it highlights minimizing unnecessary words. So how might we measure the efficiency of communications? Consider the following two examples, which endeavor to communicate the same thing (Murphy and Peck 1976).

> Example 1: "Will you ship to us any time during the month of October or even November if you are rushed, for November would suit us just as well—in fact, a little bit better—three hundred of the regular three-inch-by-fifteen-inch blue felt armbands with white sewn letters in the center? Thank you for sending these along to us by parcel post and not express as express is too stiff in price. Parcel post will be much cheaper. We are [frugal]."
>
> Example 2: "Please ship parcel post before the end of November three hundred regular three-inch-by-fifteen-inch blue felt armbands with white sewn letters in the center.

The previous correspondence (example 1, which uses eighty-three words) was a written communication from a business executive to a firm the company has dealt with for five years. You will note that example 2 (which only uses twenty-seven words) conveys the very same message, but it does so much more efficiently.

A closer look at the two examples reveals that example 2 uses just over 30 percent of the words used in example 1. Thus, it could be argued that in many cases an efficient communication can be made leveraging only a few critical words. Said differently, 20 to 30 percent of words convey the desired message. If you think that this sounds familiar, you are on the right track. Dr. J. M. Juran (1995) explained the Pareto principle, commonly called the 80/20 rule, which resulted from his realization that relatively few factors influence a final effect. Dr. Juran also referred to this relationship as "the vital few and the useful many." Applying this to our discussion, as a general rule, you should ensure that your use of words follows this approach—focus on the vital few words and shun the useful many. Murphy and Peck (1976) also offers a list of commonly used words that could be converted to be more concise (efficient) ones as shown in table 9.2-2.

Table 9.2-2: Common Phrases

Wordy	Concise (efficient)
A long period of time	A long time
Along the line of (salary)	About (salary)
At this time	Now
Due to the fact that	Because
Endorse on the back of this check	Endorse this check
From the point of view of	As
In accordance with your request	As you requested
In the city of Chicago	In Chicago
In the neighborhood of	About

The importance of effective and efficient communication cannot be overstated. Indeed, "communication is the heart and soul of management at every level. Without communication skills you can't express your thoughts, convince others, or negotiate. With good communication skills you can expect to win arguments, make sales, mediate, educate, inform, and even inspire" (Shipside 2007).

We return to Josef who is now aware of the importance of effective and efficient communications. He also recognizes the adage "When in Rome, do as the Romans do." Thus, he realizes that how and what he communicates to his engineering colleagues will likely be very different from the same information shared with those lacking engineering expertise. This is of particular concern for Josef given his aspiration to move into the production area.

A few days have passed since the delivery of Josef's performance appraisal. He is sitting at his desk, studying the circuit listed in figure 9.2-3 that we discussed in the opening chapter.

Figure 9.2-3: Transistor Switching Circuit

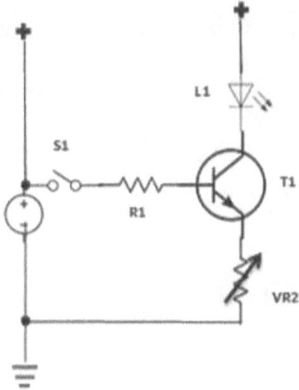

Recall that the description of functionality provided earlier was as follows:

The basic switching circuit is enabled (turned on) when switch S1 is closed thereby supplying a positive voltage (and current) to the base of transistor T1 through current limiting resistor R1. The light emitting diode (LED or L1) is then illuminated as current flows through the transistor collector and emitter, then through variable resistor (VR2) to ground. If the LED is not as bright as desired, VR2 may be varied (reducing resistance and thus increasing current flow) which will, in turn, increase the brilliance of the L1. Conversely, VR2 may be adjusted in the opposite direction to reduce L1 brilliance.

Maria Summers (regional product portfolio director) happens to walk by Josef's desk and observes him studying the circuit and asks what he is doing. Josef thinks to himself, *This would be a great opportunity to test-drive my communications skills.* He then looks at Maria and says, "Think of this switch (pointing to the S1 component on the diagram) as a water faucet. When it is in the down position, water flows through these components and gets to here (pointing to the L1 component on the diagram). Now instead of water, this is actually electric current, and when the current gets to this point, the light comes on."

Maria responds, "Oh, I see." She then asks, "What are these other components?"

Josef says, "Just the power supply, some resistors, and a transistor."

Maria smiles and says, "Got it. I thought this stuff was complicated." She then says, "Well, got to run. Thanks for the tutorial. I feel like a pro now."

While there are any number of ways in which to explain circuit functionality, the key point is that efficient and effective communication is always the best place to start.

9.3: Managing Multiple Priorities Concurrently

Returning to Josef's feedback session, recall that Daniela also expressed concerns about slippage of the new equipment installation schedule. In my book titled *All the Way to the Top: A Practical Guide for Corporate and Business Leadership*, I discuss a method for continuous improvement that I refer to as "the 5C Leadership Improvement Model" (5CLIM) or simply the 5Cs (conversation, calculation, collaboration, communication, and cognizance). I encourage you to review the entire model as the bulk of this discussion will be limited to the fifth C—cognizance. The descriptive model is shown in figure 9.3-1.

Figure 9.3-1: The 5CLIM

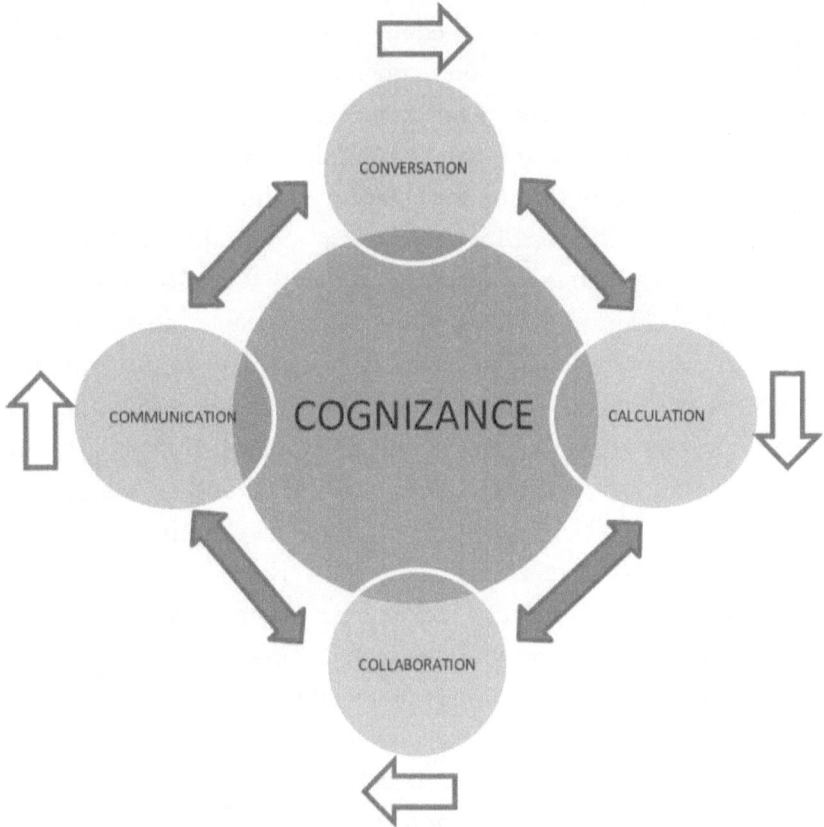

Merriam-Webster (2007) defines cognizance as "apprehension by the mind ... awareness ... notice, heed ... particular knowledge ... conscious recognition." Despite the general clockwise progression of the model and bidirectional stage-to-stage interactions, cognizance must be present in all stages of the model as reflected graphically in figure 9.3-1. Its influence is not circumscribed by any particular stage. Being aware of communication style is imperative, and awareness is a key defining factor of cognizance. However, in the context of the 5CLIM, cognizance also suggests a keen awareness of the efficacy of executing any of the core stages.

Let's say that the message output from the communication stage comprised the annual operations improvement plan. Let's further suppose that key goals, objectives, strategies, milestones, and metrics were included in that plan. Is it enough to simply assume that despite the resonance of the connection made during the communication, all key milestones and metrics will be met and that no further inquiry is required on the part of the communicator? The shrewd leader recognizes the need for implementation of incremental awareness infrastructure in the form of monitoring and measures that enable the tracking of progress toward improvement. While there are many convenient mechanisms to accomplish this, one simple and efficient way of monitoring is a dashboard review. The best analogy for this approach is a vehicle dashboard. When you're driving, there is perhaps nothing more important than focusing on the road ahead. But many things vie for our time and attention during our time behind the wheel—our passengers; the vehicles in front of us, beside us, and behind us; and more. So as drivers, we need to know certain critical information about our vehicles, such as how much fuel we have, the speed we are traveling, and the engine temperature while concurrently maintaining focus on our surroundings. This information is provided to us in a concise manner by our dashboard. Each time we glance at it, we implement a dashboard review. Compared to some earlier-model vehicles, contemporary vehicle dashboards include much more information than simply fuel level or battery charge.

Similarly, in an organization the dashboard review can take on as much or as little detail as necessary, which may also be a function of the audience for that dashboard. A CEO, for example, might be interested in the company sales, manufacturing costs, policy compliance, and public relations, while a project manager may be interested in knowing the progress of multiple ongoing construction efforts.

As an example of the use of dashboard infrastructure for tracking, assume the following metrics were communicated in a department meeting focusing on the upcoming year:

- Reduce overall operating supply orders by 10 percent by the end of the first quarter.
- Reduce inventory reporting errors by 20 percent by the second quarter.
- Improve sales in region Z by 15 percent by the third quarter.
- Procure, install, and transition to new PCs for all employees by the fourth quarter.

Here we have two dimensions to be aware of as the year progresses. We have quantifiable requirements of N percent, and we have time-driven milestones. What we want to do here is have some method to understand the progress or lack thereof at a glance and to provide any necessary information to the stakeholders for the deliverables that they believe to be of concern. Let's concentrate on creating a tracking mechanism for these items. We'll use a traffic light to represent progress like this:

- [R] (Red): Significant issues have surfaced that render this deliverable unachievable.
- [Y] (Yellow): Issues are on the horizon that could impact the deliverable if left unaddressed.
- [G] (Green): All aspects of the deliverable are on track. No issues have been identified.

The terminology used for each of the traffic light colors will vary from textbook to textbook, and it certainly isn't my aim here to reconcile the colored indicator descriptors. Likewise, those fluent in project management may recognize this approach to be similar to the standard monitoring infrastructure. The application of purest project management techniques would also include risk analysis and mitigation strategies and would focus on the relationship between scope, time, and cost. It is beyond the scope of this book to address project management components in detail. Additional inquiry into this area, if desired, is left to you.

Let us now consider how the objectives we previously discussed might be presented in a dashboard view through the use of the traffic indicators. Please refer to figure 9.3-2 for this discussion.

Figure 9.3-2: Dashboard Simulation

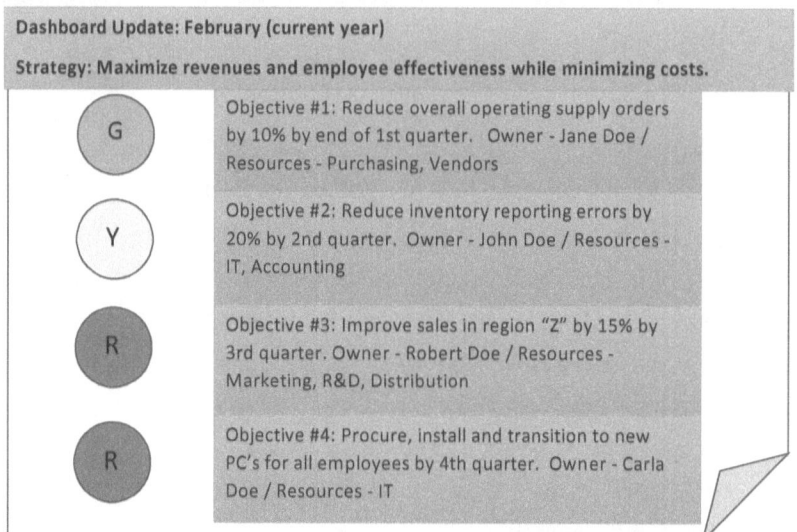

Note that the dashboard has a heading labeled "Strategy." So what's up with that? "Although business strategy is fairly new, many of its concepts and theories have their antecedents in military strategy, which extends back to principles enunciated by Julius Caesar and Alexander the Great" (Grant 1995). The strategy listed previously, namely to maximize revenues and employee effectiveness while minimizing costs, may be thought of as a means for channeling resources to execute broader business goals. Chandler, as cited by Grant (1995), shared this perspective, suggesting that a strategy is "the determination of the long run goals and objectives of an enterprise, and the adoption of courses of action and the allocation of resources necessary for carrying out these goals."

Notice also the reference to resources included in the dashboard. These are the departments and areas that will be required to assist in the completion of the objective. Choosing the right resources is a critical step in the dashboard review because it allows the key stakeholders to recognize their roles in strategy execution. Focusing specifically on the internal aspects of the dashboard, we can quickly observe that all the objectives fall under a much broader strategy (maximize revenues and employee effectiveness while minimizing costs). Objective #1 is on track with no issues identified. Objective #2 will require some intervention in order to avoid missing the target completion date. Regarding objectives #3 and #4, significant issues have surfaced that render these deliverables unachievable.

Also notice that in addition to the specific objective, an owner has been included. This is extremely important for at least a couple of reasons. First, by allocating a name to the specific objective, the stakeholders are now able to engage with the owner in hopes of removing barriers or offering suggestions that may have otherwise been overlooked. Second, assuming that the owner was made aware of the dashboard review process, it ensures that the owner had assessed all angles and possible solutions before designation of the objective status and now welcomes assistance and input from others.

Typically, a dashboard review is presented to key stakeholders within the organization with the intent of not only notifying but also garnering support for the allocation of additional resources (people, dollars, and time). On objective #3, the allocation of people could include interim increases to the sales force such that retailer calls are increased. Likewise, dollar increases might be allocated in support of increased "buy two get one" programs aimed at building product trial in hopes of increasing longer-term product usage. The reference to time is in the context of potentially extending the deliverable date, in this case allowing Carla to shift the due date back to the first quarter of the next year.

Often the dashboard is completed by a resource designated within a department or by a central departmental resource within an organization based on inputs received from the deliverable's owner—the owner being the person responsible for the line item. All updates are supplied by the owner and are issued to the compiling person or area along with some sort of explanation (not shown in this example) of what factors are driving any deviation from green. It is also common for the owner of any deliverables or objectives whose status is not green to comment in the review forum about the specifics listed in the status area (if included) and to offer suggestions on how to mitigate the deviation. It is also likely and strongly encouraged that the owner of any objectives that are not green will forewarn key stakeholders, informing them of the issues before the dashboard review so they won't be surprised during the meeting. (As a side note, while serving as director of strategic planning during my time in the corporate sector, I was responsible for presenting aggregated departmental dashboard information to the entire operations business team. This was one of the best jobs that I ever had. And although it had very little to do with engineering, it did allow me to make the connection between business and strategy—a key enabler for higher-level leadership success.)

The dashboard is a powerful and robust tool for initiating the communication of key strategies, objectives, and milestones. Although it is presented here in the context of a much larger stakeholder environment, someone operating in the role of an individual contributor will also find the dashboard useful, particularly when the responsibility for multiple projects or priorities rests with a single person.

Consider, for instance, a procurement analyst who is collaborating with several vendors in an attempt to lower supply costs. Each of the objectives in the example dashboard could be supplanted by supplier targets for raw materials cost reductions. A quick analysis of the completed dashboard would facilitate necessary conversations, potentially resulting in new calculations, thereby driving the need for collaboration.

The frequency with which a dashboard review should occur is dependent on several factors, including the mean time between objective deliverable dates, the specific milestones associated with a certain objective, and the availability of key stakeholders to review the data. As an example, if a deliverable due date is one year from the current time, depending on the severity or magnitude of the deliverable, reviews implemented less than a monthly frequency may not allow sufficient time for key milestones to come to fruition. Likewise, if key stakeholders are available only quarterly, then the dashboard review frequency may be protracted accordingly, particularly in the case with long-term deliverables.

Now being generally aware of the cognizance stage of the 5CLIM, Josef reaches out to his friend and colleague Jessica Wright, the manufacturing director, to gain more of an applications perspective.

In the meeting with Jessica, Josef recognized that in addition to the dashboard constructed to monitor certain critical objectives with defined time lines, Jessica reiterated the need to implement a dashboard she could refer to on an ongoing basis. She referred to this as her personal operations-wide dashboard. The key question that she had to address before its implementation was as follows: What were the critical manufacturing operations metrics that may or may not have been addressed in her objectives-based dashboard? Considering all the data and processes for which she was responsible as the director of manufacturing operations, Jessica landed on the dashboard shown in figure 9.3-3.

Figure 9.3-3: Jessica's Personal Operations Dashboard

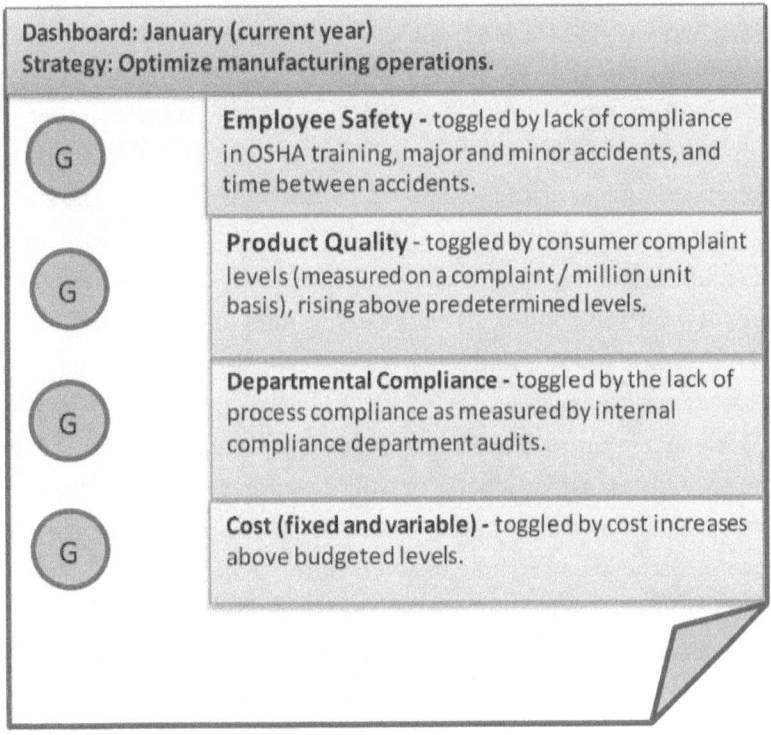

Note that a strategy is listed as was the case with the objectives-based dashboard. Here, however, Jessica focuses on what she believes are the critical ongoing operations metrics. Jessica recognized that the responsibility for the safety of every employee working in the manufacturing operations area rests with her. Consequently, she listed this as her top dashboard item. Next, recalling the issues with product quality as identified by the QA manager, she included product quality as a key dashboard item. Similarly, because Summit Consumables is regulated by the FDA, it is extremely important that all employees correctly follow existing compliance protocols. So this was a critical metric as well. The last one listed is in the area of cost incurred in operations, both fixed and variable.

Jessica also contemplated the criteria she should use to switch (toggle) the status of those metrics from green to yellow to red. Recall that toggling criteria for a dashboard may vary widely depending on the application. Despite the criteria chosen, those affected by the status changes (e.g., direct reports) must also have a good understanding of the toggling criteria.

Beginning with employee safety, Jessica recalled that the Occupational Health and Safety Organization (OSHA) mandates that certain aspects of the workplace conform to federal standards. In particular, she recognized that OSHA's mission, which was based on the Occupational Safety and Health Act of 1970, is "to assure safe and healthful working conditions for working men and women by setting and enforcing standards and by providing training, outreach, education and assistance" (osha.gov).

With this in mind, she decided that any failure to complete OSHA-required training as per schedules would switch the metric from green to red as would any major injury, such as on-the-job dismemberment. Likewise, if the average time between minor accidents (such as minor strains) decreased to a certain level, that, too, would switch the metric from green to red. Toggling from green to yellow would result from the number of minor accidents occurring regardless of the average time between occurrences.

Concentrating on quality for the moment, assume the historical product complaint rate is 320 complaints per million units. In other words, for every one million units produced, a consumer complains about 320 of those units, meaning the complaint rate is 0.032 percent. If the current complaint rate were, say, 1.2 percent, this level would far exceed the historical rate, thereby forcing the metric to red. Jessica believes that switching to yellow would occur as long as the complaint rate is less than historical but greater than the rate that is believed to be achievable. Jessica views the compliance metric slightly different. She believes that its switching mechanism would be driven by employee performance as measured by process audits. The processes that are to be audited by the internal compliance audit team are those aimed at ensuring product quality and integrity. In the event that all process audits identify no lack of compliance with process execution, then the metric would remain green. Otherwise, it will switch to red or yellow based on the level of noncompliance. (Less than 100 percent but greater than 95 percent would switch from green to yellow, and less than 95 percent would switch to red.) Jessica listed cost as the final metric on her personal dashboard. And despite the importance of fixed and variable cost, she listed it last to send the message that reduced costs must begin with stellar performance in employee safety, product quality, and process compliance. For consistency, however, she set a threshold of a 2.5 percent aggregate fixed and variable deviation from target to switch from green to yellow and greater than 2.5 percent aggregate deviation to switch to red. An alternative construction of Jessica's personal dashboard is shown in figure 9.3-4.

Figure 9.3-4: Alternative Personal Dashboard

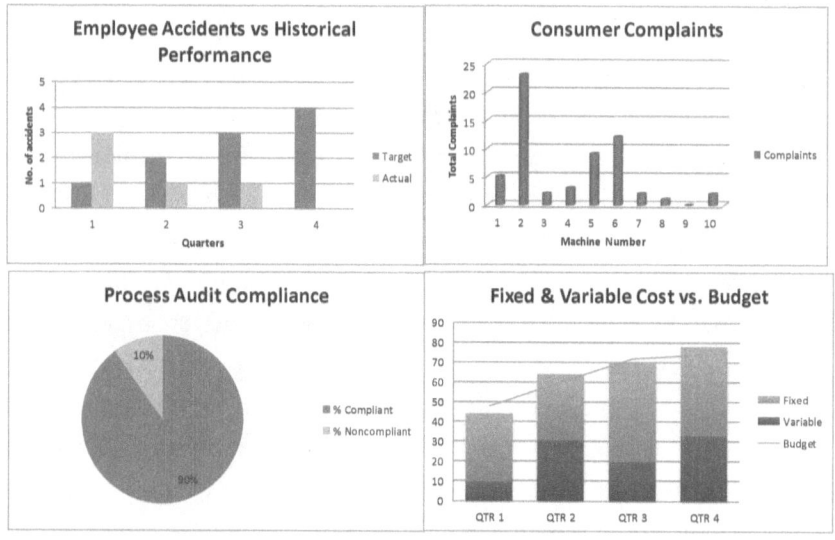

While the same critical metrics are listed, the manner in which the data suggests opportunities (the need for improvement) is shown graphically. In this case, Jessica must take a moment to read the specific data points in the charts to determine metric status. The traffic light indicators are a more efficient mechanism by which status may be determined, but the data shown in figure 9.3-4 offers a more comprehensive representation of that same information. Josef now has a good understanding of how to keep track of his key priorities to include reflecting his specific goals and objectives related to the equipment installation schedule.

As a side note to the reader, it is always a good idea to know how you are being measured. What is your boss focusing on as a means to judge and reward your accomplishments as reflected in the next performance review? Once you know how you are being measured, which might begin with a conversation (the first of the 5Cs by the way), you should develop your own personal dashboard. It is best that you know when an item for which you are responsible is going to switch from green to yellow or red before your boss brings it to your attention.

9.4: Josef's Expert Power in Action

Recall our discussion of expert power in chapter 3. As pointed out by Daniela, problem solving was one of Josef's greatest assets. And inasmuch as he was perceived to problem-solve better than just about anyone, it was also his expert power. Although only one form of power, it is indeed very powerful, and leveraged correctly, this can facilitate one's ability to influence others, which is a great indicator of leadership capacity.

We are going to fast-forward the clock several months as we check in with Josef. The equipment installation schedule is back on track, and Josef has made tremendous progress with his communication style. A significant manufacturing issue has surfaced, Josef is eager to solve it. A simplified version of the manufacturing flow is listed in figure 9.4-1.

Figure 9.4-1: Manufacturing Flow

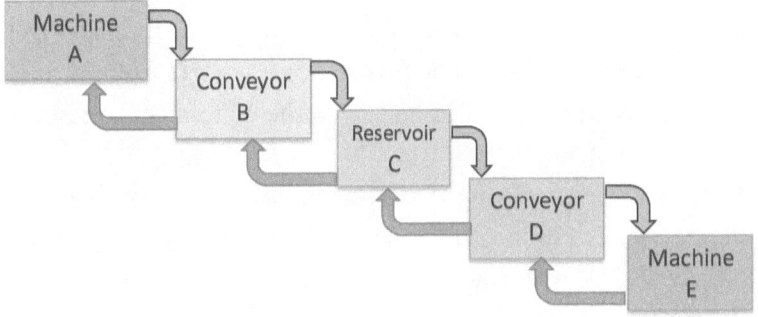

At a high level, machine A produces widgets that travel through conveyor B in route to reservoir C. From reservoir C, the widgets flow via conveyor D to machine E, where several widgets are bundled together and wrapped in the final packaging. You will also note that each of the process stage components provides feedback to the predecessor component. For example, reservoir C provides feedback to conveyor B. The aim here is to allow both conveyor B and machine A to be shut down when reservoir C reaches capacity or if conveyor D should cease to operate for any reason. Likewise, machine E would create an upstream shutdown of conveyor D in the event it ceased operating.

The problem with the manufacturing process is that reservoir C appears to oscillate (moving backward and forwards) in response to the signals provided from conveyor B. As a consequence, the belts on the reservoir break under the constant stress, causing the equipment to shut down and resulting in tremendous downtime for the entire production line. The original equipment manufacturer is aware of the issue and communicates that a fix has been developed, the cost of which is more than $30,000 for each of the fifteen machines installed. In addition, the fix would require substantial equipment downtime, resulting in a tremendous impact to production. Josef studies the problem and develops a simple hardware solution that he believes could be implemented during the scheduled machine downtime. However, he did point out that he would need access to the manufacturing equipment to test his proposed solution. When asked by the production manager what specifically he planned to do, Josef responded as follows: "I am going to smooth out the signal provided from machine A to conveyor B, resulting in little to no reservoir oscillation." The production manager thought the idea sounded great and granted him access to the equipment to test out his proposed solution.

Back in the engineering department, Josef was asked to provide an update to his engineering manager, who was also an electrical engineer with a degree. Josef responded by sharing a process diagram and describing the problem. He then talked about the existing control circuitry (software and hardware) and explained how he planned to develop a simple PID (proportional-integral-derivative) controller that would be attached in the main controller cabinet.

Fast-forwarding a bit once more, Josef's solution resolved the issue during the pilot test and was subsequently installed on all machines, resulting in the total elimination of the issue. In turn, this significantly improved uptime, decreased material costs for repairs, and allowed for the reallocation of technician resources to other production issues. The total equipment and installation cost for Josef's solution was less than $600 per machine. Thus, a cost avoidance of over $400,000 was realized when considering the OEM proposal.

Within just a few months, Josef, employing excellent communication skills and staying on top of all the details associated with his assignments, solved several other manufacturing issues. He actually became quite the go-to person.

CHAPTER 10

Josef's Big Break

RESULTING FROM JOSEF'S success with the equipment installation and problem resolution, his name was being circulated for a possible group leader position in the area of quality engineering. When Josef first learned of the interest in him for this area, he approached his director, Daniela, who confirmed the rumors and asked how he felt about taking a position that was not exactly what he may have previously expressed interest in doing. Josef reasoned that while this was indeed a different start to the career path that he had envisioned for himself, opportunities to move into a defined leadership position were few, and qualified candidates were many. With this, Daniela informed Josef that the need was immediate and that he would be transferring within the week. Figure 10-1 reflects Josef's new organization structure.

Figure 10-1: Quality Engineering Organization Structure

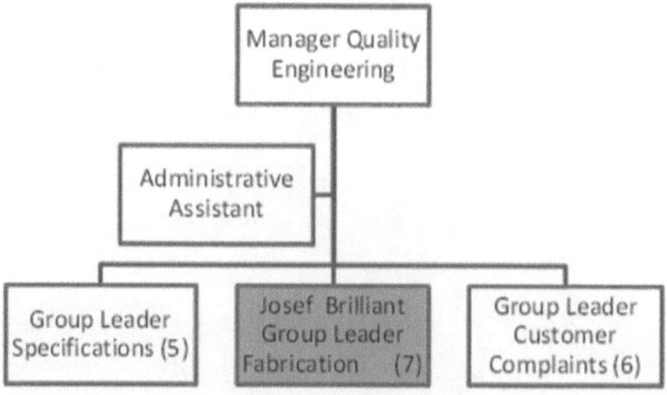

As indicated, Josef has two peers (Mike and Susan) and seven direct reports. The manager of quality engineering (Don), like Josef, has an engineering undergraduate degree. Unlike Josef, however, the manager (Don) graduated number two in his class from a top-ten engineering school with a degree in mechanical engineering. The manager used this well-respected status to sort of banter with Josef from time to time. Josef welcomed this approach and oftentimes sought to draw comparisons in a playful way. In one such instance, Josef stated that "those who received degrees in mechanical engineering only did so because they could not make it in electrical engineering." Don responded, "An electrical engineer's day at the beach consists of sand, waves, and flip-flops."

Josef had long believed that of the engineering disciplines, mechanical engineers were most likely to succeed and progress in leadership. Josef's peers Mike and Susan have degrees in phycology and business respectively.

In his current position, the role of Josef's team is to provide statistically oriented engineering support to the production area. This includes developing testing protocol for the evaluation of equipment (e.g., product quality, material qualifications, impact to utilization of machine modifications, efficiency, etc.). The education level of Josef's staff ranged from degreed engineers to biology majors to one employee who held an undergraduate and graduate business degree and one employee possessing a two-year associate's degree. In his role as electrical engineer, Josef was accustomed to identifying and solving problems employing novel approaches. The manager of quality engineering believed that this sort of thinking and approach would add tremendous value and credibility to the department and also build employee capability.

Managed by the human resources department, Summit Consumables implements a process referred to as "new manager assimilation" for all first-time leaders holding permanent positions on the company's organization chart. Josef meets the criteria for this process. In general, the process mirrors a 360-degree feedback session and requires the leader undergoing the process to meet with his or her direct reports to discuss the feedback. On behalf of Josef, HR issued the MLQ as was discussed in chapter 6. Listed in figure 10-2 are the results of Josef's quantitative MLQ feedback.

Figure 10-2: Josef's Feedback

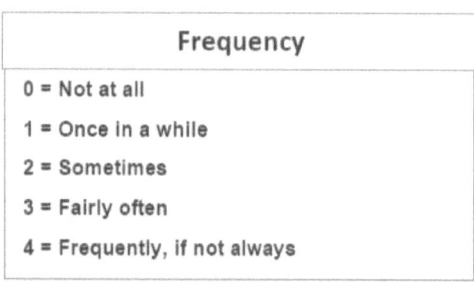

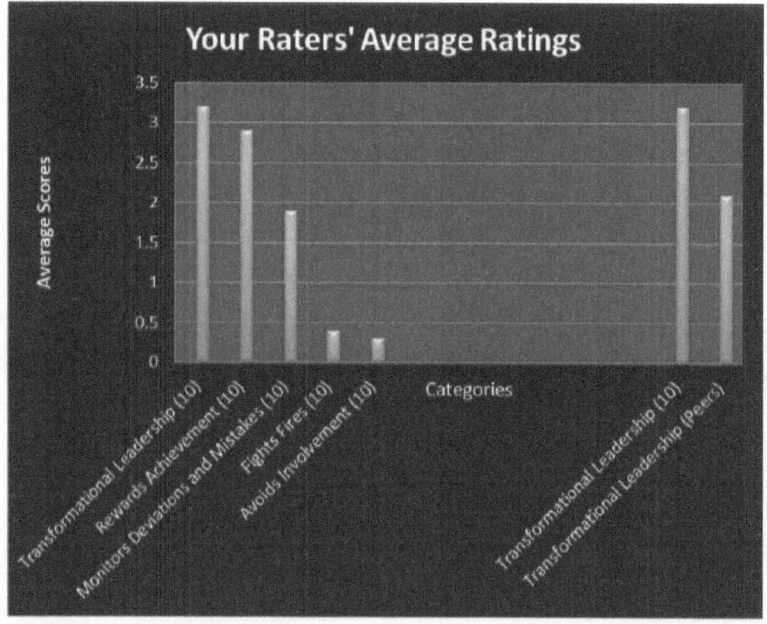

Upon review of figure 10-2, you will note the table labeled frequency. This simply refers to the frequency of observation of a specific attribute. These attributes are then totaled and averaged for each of the full-range leadership categories as exhibited by the sample forms presented in appendences B and C. The bar chart reflects the combined results that Josef received from his boss, peers, and subordinates. Beginning with the score of 3.2 out of four in TL, this represents a significant accomplishment for Josef given that this is his first formal leadership position held. It is also critical to recall that "transformational leadership is a process of influencing in which leaders change their associates' awareness of what is important and move them to see themselves and the opportunities and challenges of their environment in a new way" (appendix E). Josef's score here is closely followed by a score of 2.9 in the XL component (contingent reward and rewards achievement). Recall that contingent reward is one aspect of XL and that "transactional leaders display behaviors associated with constructive and corrective transactions" (appendix E). Recalling our previous discussion, it is not at all uncommon for a leader to exhibit the transactional component of contingent reward. Indeed, it is incumbent upon leaders to provide direction regarding what is to be accomplished. You will note, however, that in the second transactional component (management by exception active - MBEA), Josef scored much lower at 1.9. Referring to the normative tables in appendix D, the score of 1.9 is closely associated with the sixtieth percentile in two of the three normative tables and between 1.75 and 2.0 in the third one. This suggests that Josef's actions in this area are higher than 60 percent of those tested. It should be noted that unlike the TL and the XL component of CR, a lower score in the area of MBEA is preferred. It is not at all uncommon for new or relatively inexperienced managers to score higher in this particular area. After all, this is their first attempt to prove themselves in leadership, and it is quite normal for them to take a more hands-on approach to leadership. The remaining two leadership categories (firefighting or management-by-exception passive [MBEP] and avoids involvement – laissez-faire) revealed scores for Josef as 0.4 and 0.3 respectfully. The first of these two categories (firefighting) reflects a leadership style that "does not respond to situations and problems systematically" (appendix E). In stark contrast to the CR component of XL, leaders exhibiting MBEP "avoid specifying agreements, clarifying expectations, and providing goals and standards to be achieved" (appendix E). The final category, laissez-faire (LF) is, in essence, no leadership. Josef's scores in MBEP and LF mirror the low end of the percentile scoring range reflected in the normative tables found in appendix D. Here again, lower scores are better in these areas. The bars on the far right reflect how Josef scored relative to his peers who previously underwent MLQ assessment. Consistent with the research discussion, you will note that Josef scored higher in TL than his peers. Please recall the five constituent elements of TL (IIA, IIB, IM, IS, and IC). Although not included with Josef's summarized feedback, his scoring in the TL category of intellectual stimulation (IS) was at a 3.9, which in all the normative tables represents a roughly 95 percentile range. This, too, should not be surprising when considering the previously discussed research. You will recall that when the larger populations were tested (managers and the integrated population), the null hypothesis stating that there was no difference in the perceived style of IS between those with engineering degrees versus those without was rejected, suggesting that those with engineering degrees were perceived to demonstrate more of the IS style as evidenced by the two-tailed Sig values.

There is a second component of feedback that is provided with the MLQ, specifically qualitative feedback. Here, the raters have the opportunity to respond to open-ended questions regarding the person being rated. The results of these responses to customized questions pertaining to Josef are shown in figure 10-3.

Figure 10-3: Supplemental Assessment Questions

Supplemental Subordinate Assessment Questions:

1. What are two or three things that would help this person be more efficient?
   Share the vision
   Conduct weekly staff meetings
   Take time to listen to our ideas

2. One thing that gets in the way of this person's effectiveness is:
   Allow us to join in the problem solving process
   Lack of engagement of team members

3. What I admire most about this person's leadership:
   Very intelligent
   Adjusts communication style to audience

The first two questions address how Josef might maximize his leadership efforts. As reflected in appendix E regarding the TL constituent individualized consideration (IC), "transformational leaders pay attention to each individual's need for achievement and growth by acting as coaches or mentors. Followers are developed to successfully higher levels of potential. New learning opportunities are created along with a climate in which to grow." Again, although not shown, IC was an area where Josef scored the lowest, coming in at 2.1. Referring once more to the normative tables, the best case table value for this range (found in the lower level ratings table) is at the twentieth percentile. Both the peer and higher-level ratings were around the tenth percentile for this attribute. So what might this mean? Let's revisit the responses to the first two questions. It could be that each of the responses reflects opportunities for Josef to develop others as is subsumed by exhibition of IC. Despite Josef being rated extremely high in the area of IS, referring specifically to the feedback received from question #2, he does not seek to build this same capability in others. Perhaps Bonasso (2001) said it best when he wrote, "Leaders are risk-taking educators who teach us how to learn from our mistakes by defining today's failure as the information base for tomorrow's success." Further to this point and as indicated in figure 10-4, Compton-Young and others stratified the results of a survey reflecting "the leadership capabilities perceived important by professional engineers ... prioritized from 1-3, where 1 is the most important and 3 is the least important."

Figure 10-4: Leadership Capability Survey Results (Compton-Young et al. 2010)

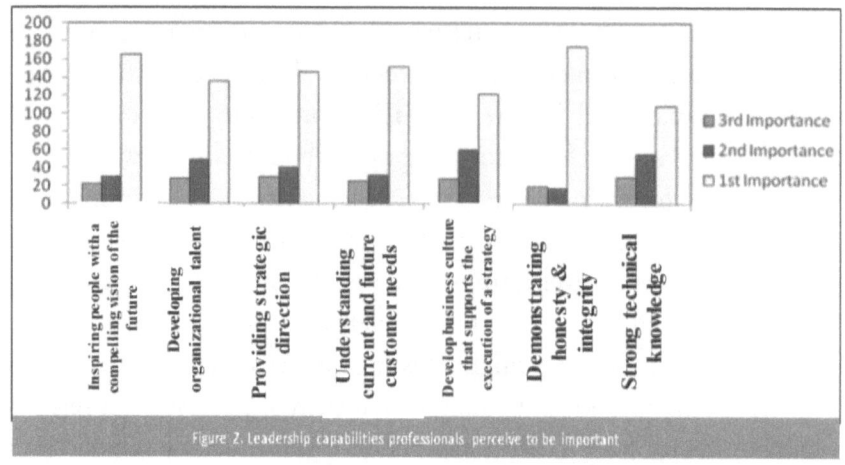

I call your focus to the items listed as "First Importance" (also the tallest bar). Of the top seven first-importance items, it could be argued that five of them directly relate to the supplemental feedback received by Josef and are also linked to the TL constituent IC. You may be wondering what the fourth item from the left, "understanding current and future customer needs," has to do with IC. It is first important to understand that the term *customer* does not simply refer to those who purchase products, goods, or services from a company or business. As a leader, you have customers on a 360-degree basis (subordinates, peers, and bosses) to whom you are a supplier. And as the supplier, your product (output) can be measured and thus improved. To better understand this perspective, consider the fundamental systems view of inputs, processing, and outputs as was previously introduced in chapter 2.

On a micro level, consider that every response or action—well, almost every response or action that you execute—requires some level of input that is subsequently processed prior to drawing a conclusion or acting on the input. Indeed, as you are reading this paragraph, you are also taking in (inputting) information and subsequently processing that information. There's no doubt you will formulate an opinion or output regarding that information. I hedge here with the word *almost* to recognize the will and capacity of research pundits' attempts to find some esoteric exception to this model and also to recognize that despite your ontological position, very little in this world can be assumed to be 100 percent applicable in every situation. Be that as it may, I now invite you to think of a situation to which this model (input, processing, and then output) is not directly applicable. Keep in mind, however, that as you are contemplating a scenario for which this model is not applicable, you are searching your brain's database (as simulated in figure 10-5) for an IPO scenario that may not align with this model. Your respective consideration of the front end of the model (input) and the effective processing of that input will undoubtedly result in your determination as to the efficacy of this model to that particular situation (output). And in doing so, all stages of the IPO model are implemented.

Figure 10-5: IPO Model—In Search of a Contrary Activity

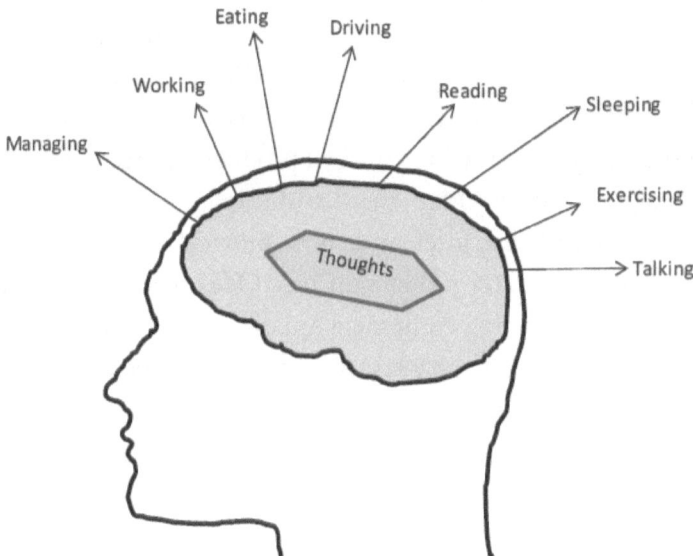

Not there yet? Here is a quiz question for you: Considering the most recent presidential election (whereby the population votes for the candidate of choice), how would you apply the IPO model on a macro basis? One simple way to do this would be to consider the votes as inputs to the voting machines in aggregate. Then you need simply recognize that as a result of counting the votes (whether completed by humans or machines), the processing of the votes yields an output that is, of course, the next elected president.

Okay, you probably want to know what the heck this has to do with Josef and his customer base. Great question! Being responsible for the development of his direct reports, Josef receives input from them from various sources (e.g., their progress on assignments). He then processes that input (e.g., determines if the progress is consistent with milestone expectations). This is followed by some output (e.g., providing direction, coaching, support, etc.). Thus, the entire systems model is implemented. You will note that Josef's output may take on several different forms. The driver for these varying scenarios is the particular situation or circumstance. This contingent leadership approach was discussed in chapter 2 and will be revisited in the next chapter.

Inasmuch as Josef applying the IPO model to the development of his direct reports represents a micro example, a macro example is simply an expansion of the process. Said differently, think of a simple manufacturing process whereby raw materials are procured (inputs) and subsequently converted to finished products (processing) and sent to finished goods (output) for later distribution.

Likewise, recall that Josef's role is to provide statistically oriented engineering support to the production area. One aspect of role execution is to receive requests for equipment qualifications (inputs), convert those requests in to actionable test plan designs (processing), and then implement the test plans in the manufacturing environment, resulting in a final report to customers (output). The micro component here would be Josef's review of the specific test plan designed by an assigned person in his group to determine whether or not he concurs with the proposed plan approach.

The value of the IPO model cannot be overstated. Take a moment to consider your specific role within your organization. Whether you are a leader or a sole contributor, you still have a role that involves you receiving some sort of input, acting on that input in some way (processing), and subsequently providing an output. Now the key thing here is that if you simply tie in the feedback loop that we discussed in chapter 2, you can improve the quality of your output by simply addressing one or both of the other two variables (input and/or processing).

Returning to the example previously provided, let's say that the test plan that Josef's team implements has to do with assessing whether or not a certain modification to the manufacturing equipment actually impacts product quality. Let's further assume that the design engineer is extremely confident that the product quality will be improved, but upon completion of data analysis, no statistically significant difference is detected in product quality with or without the modification. Upon Josef's review of the original test design implementation (processing), he noticed that two separate machine operators were utilized but that they each had different machinery settings. Consequently, he learned that when the design engineer initially tested his modification, he did not consider the effect of varying operator settings. Thus, the original request (input) was not clearly defined. With this in mind, the design engineer made a slight adjustment to the original equipment modification, the test plan was rerun, and the expected improvement was observed on a statistically significant basis. Now one might argue that the IPO model requiring adjustment was that of the design engineer's and not Josef's. However, in the end, Josef's review resulted in the best outcome for the company. And after all, if one area of the organization succeeds, then we all succeed, right?

On a personal note, viewing what you do in a systems way (IPO) will no doubt yield tremendous returns. Thus, I invite you to consider your role (or that of your department) and seek to improve upon your outputs.

CHAPTER 11

Leading from Behind

IN THE PREVIOUS chapter, we identified several ways (contingencies) in which Josef might respond to his team (e.g., providing direction, coaching, supporting, etc.) in the context of development. Here we will explore various applications of this approach. When most people think of the word *leader*, they imagine the scenario whereby the person in charge is directing the actions of his or her direct reports. Recall our discussion in chapter 3 regarding legitimate power. However, this is only one aspect of leadership. And insomuch as leadership is about contingent actions, highly effective leaders take the TL constituent of individualized consideration (IC) one step further and actually view themselves as serving those being led. Yep, you heard correctly. Despite the leadership contention to the contrary, it is not simply the leader's role to provide direction (tactical or strategic) and to leave it to the team to recognize their limitations and *fix* them. While it is certainly a good idea to empower the team, doing so without seeking to identify and address individual member shortcomings (development opportunities) is a recipe for disaster. Northouse (2013) cited views of how the leader might serve the needs of his or her team—listening, empathy, awareness, persuasion, conceptualization, and commitment to the growth of people.

Although we have touched on many of these servant-oriented attributes in prior chapters, it is important to recognize the tremendous value added by employing them in the context of team development. In chapter 3, I introduced you to the 5C Leadership Improvement Model (5CLIM), which I cover in detail in another tile, but I only discussed the fifth C (cognizance) with any level of detail in this book. The first of the 5Cs (conversation) is a key enabler for serving the team. Through implementation of healthy conversation—including listening—the leader is able to develop a keen understanding of individual strengths and development opportunities. With this in mind, let us revisit Josef's team members. You will recall that education level of Josef's staff of seven ranged from engineers with degrees to biology majors to one employee who held an undergraduate and graduate business degree and another employee with an associate's degree. More specifically, the team member names and education are as follows: Tawanda (MBA), Karla (BS business), Harold (BS biology), William (BSME), Gerald (BSEE), Dana (BS statistics), and Douglas (AAS technology).

Ahead of discussing each of Josef's team members, it is important to understand how goals and objectives should be set. To this end, according to Armstrong (2009), "Many organizations use the SMART mnemonic to summarize the desirable characteristics of an objective [where the acronym is delineated below]

S = Specific/stretching—clear, unambiguous, straight forward, understandable and challenging
M = Measurable—quantity, quality, time, money.
A = Achievable—challenging but within the reach of a competent and committed person
R = Relevant—relevant to the objectives of the organization so that the goal of the individual is aligned to corporate goals
T = Time-framed—to be completed within an agreed upon timescale."

Just as a side note, some scholars believe that the M should reference *motivating* instead of *measurable*. The argument here is that in order for a goal to be specific, it also has to be measurable, and therefore, the use of measurable in the acronym is redundant. Be that as it may, the key takeaway here is that goals or objectives should have the appropriate level of specificity, offer an achievable challenge, be quantifiable, and relate to the broader organization goals. Once the SMART goals and objectives have been developed (refer to the sample shown in chapter 9), a cognizance dashboard (similar to the one also shown in chapter 9) can then be developed.

Having worked with his team to develop SMART goals, Josef chose to implement quarterly review sessions (conversations) with each of his direct reports. Resulting from his recent conversation with Tawanda, a recent MBA graduate, Josef quickly realized that Tawanda was adept at seeing the big picture regarding her assignments, was very self-confident, had great transferrable skills, and was a great communicator. From a developmental perspective, Tawanda would often get bogged down with the details when it came to executing test and evaluation plans reflecting opportunities in the area of planning and organization. As a result, she took a bit longer in completing her assignments than some of her peers. And possibly resulting from her self-reliant approach, Tawanda missed out on opportunities to leverage and learn from her peers. Thus, for Tawanda, Josef encouraged her to consider the 80/20 rule, focusing on the major variables of a problem or solution (such as test plan development) and not to worry so much about aspects that were not material to the outcome following implementation. This approach worked well for Tawanda and served to complement her "big picture" worldview. Despite pushback from Tawanda, Josef also sought opportunities to have her team up with peers to accomplish certain assignments. In time, however, Tawanda began to appreciate the differing perspectives offered by her team members.

Karla, similar to Tawanda, exhibited good communication skills, but because of the now highly technical nature of the department's work, she appeared somewhat tentative. This was particularly evident when she met with customers representing the manufacturing engineering department aimed at developing equipment evaluation plans designed by a former group statistician. In her words, the engineers were "too technical" and made things harder than necessary to understand. Karla, who had spent the last ten years in the department and who was somewhat opposite to Tawanda, was use to approaching plan design by employing a step-by-step process approach. However, much of Karla's test plan design was predicated on templates constructed by a former group statistician—many of which were now somewhat dated. Thus, it was becoming increasingly difficult to apply this cookie-cutter approach to meet the needs of her current manufacturing engineering customers. In his conversation with Karla, Josef expressed his appreciation with the approach to standardize test plans. At the same time, he recognized that Karla lacked some key competencies regarding the functionality of the new equipment. He believed that with a good understanding of how things worked, she would be more comfortable interacting with her customers. Leveraging his relationship with Jessica Wright (manufacturing director), Josef arranged for Karla to participate in a three-week training program shadowing the equipment maintenance technicians. Karla thought not only that this was a great idea but that it also showed Josef's willingness to go the extra mile to assist in her development.

Harold can best be described as consistent. He was quite diligent in the execution and delivery of assignments, kept copious notes from staff meetings and most often was the first one to show up for work. Harold also had a penchant for talking over others which at times seemed to frustrate his peers. Additionally, Harold seemed obsessed with achieving the highest available performance rating at all cost. Consequently, he desired very specific direction regarding Josef's expectations of him. One way that this behavior manifested itself was Harold's desire to have weekly meetings with Josef just to ensure he was on the right track with all his assignments. Thus, Josef recognized the need for Harold to function in a more autonomous fashion. The question was as follows: How might Josef get Harold to see the value in this approach? Pondering this opportunity, Josef concluded that one key enabler for employee autonomy was having trust in the leader. This should not be surprising as, according to Liu and others (2010), "early studies have suggested that trust in the leader and self-efficacy are two significant mediators between TL and followers' outcomes." The cited definition of trust presented by Liu and others was "the willingness of a party to be vulnerable to the actions of another party based on the expectation that the other will perform a particular action important to the trustor, irrespective of the ability to monitor or control that other party." Thus, it was Josef's plan to instill this sort of trusting relationship with Harold. In doing so, Josef had to resist the temptation to immediately jump in to assist or provide direction to Harold at every turn. Instead the aim would be to encourage Harold to rely on himself to figure things out. At the same time, as the leader, Josef recognized that a failure by Harold was a failure by him too. With this in mind, Josef began implementing his strategy in baby steps, encouraging Harold to take more and more liberty with his work assignments over time while staying true to the objective of the assignments. Josef also provided Harold with assignments that allowed him to leverage his strengths while offering opportunities to shore up his weaknesses. Finally, Josef realized the need to shore up Harold's data analysis skills by allocating time and financial resources for him to attend a SPSS statistical software workshop. This was an important step as it evidenced Josef's willingness to take a vested interest in Harold's development.

William, much like Josef, was extremely intellectual and a quick study on the business side. Giving credence to Josef's view of mechanical engineering majors, William was pegged to succeed Josef in his current group leader role. As a point of fact, many believed that it was only a result of William's lack of engineering project management with Summit Consumables that he did not land the group leader job. Unlike his peers, William did not exhibit any real shortcomings. He was at the top of his game with his work. He had great communication skills, and his work ethic rivaled that of Harold's. Being aware of this and also seeking to circumvent any feelings of animosity that William might otherwise have, Josef sought opportunities to delegate and encourage William to leverage the inherent TL traits of intellectual stimulation (IS) to seek innovative ways to look at work processes within the department. As an example, William was asked to take a look at the existing equipment qualification process and determine how he might streamline the process. In response, William identified an innovative statistical approach, specifically sequential probability ratio testing (SPRT), to supplant existing testing and sampling methodologies. William then trained his peers as well as a peer group on its use, and he even led a training session in the engineering area. Needless to say, William received a great deal of praise and recognition for this approach as it resulted in less waste during equipment evaluations and less iteration was implemented. When asked if Josef felt a bit intimidated by William receiving all the accolades, Josef responded; "As a leader, my job is to incessantly seek to grow the skill and capability of my team in every way possible so that if I were to leave, the department could continue to flourish."

Dana, who possessed a BS in statistics, was obviously well grounded in the area and required little to no assistance executing the technical aspects of her work. However, Dana appeared to be prone to engaging in various forms of conflict. On one specific occasion, Dana and a peer department representative had a heated disagreement about how to approach problem resolution related to a customer complaint. Although it is not at all uncommon to encounter opinions that differ from those of your own, Josef recognized that conflict seemed to follow Dana within and outside of the group on a frequent basis, and he knew that inaction on his part could result in Dana's career spiraling downward. With this in mind, during their next scheduled meeting, he began the discussion by stating that whenever two or more people come together, there is the possibility of conflict. He pointed to situations where he had experienced conflict. He then discussed the five widely accepted styles of resolving conflicts originally developed by Kenneth Thomas and Ralph Kilmann in the 1970s (collaborating, competing, compromising, accommodating, and avoiding). He took this a step further by asking Dana to role-play with him, and in each role, he challenged her to identify the best conflict resolution approach. Dana found this exercise to be extremely helpful, and in time, Josef was witness to her growth in this area. The two of them often joked about the past with Dana, saying "Josef, could you be a bit more accommodating when it comes to my next pay raise?"

Josef would respond by saying, "Come on, Dana. I thought you agreed to compromise."

Douglas was the only member in Josef's group not possessing a four-year degree. He had an AAS in engineering technology. Douglas typically supported the remaining team members by collecting, sorting, and storing product samples and maintaining departmental equipment calibration records via an automated data collection system that only he could access. Recently, Josef noticed that Douglas was not showing up for the weekly staff meetings on time and was oftentimes missing them altogether. Likewise, his biweekly status reports were either not complete or missing altogether. Making matters worse, Douglas called out sick on the day of a very critical manufacturing equipment qualification test. Thus, there was no one available to collect the necessary samples, and the equipment calibration verification, which occurred prior to every evaluation, could not be completed because the team was locked out of the system after multiple failed attempts to access the software. Efforts to contact Doug were futile as he was not answering his phone and his voice mail was not set up to receive messages. The next day Douglas arrived for work, went directly into his cubicle, and began playing solitaire on his computer. Josef happened by, and noticing Douglas's actions, he invited Doug into his office for an unplanned meeting. Upon entering the office, Douglas began, "Boy, it's sure good to be back at work. I was so bored yesterday. Did anything interesting happen in my absence?" Josef brought Douglas up to speed but also explained how he could have been a bit more proactive in getting things ready for the very critical test that, as a consequence of his actions—or lack thereof—had to be rescheduled. Josef also took time to review Douglas's behavior regarding staff meeting attendance as well as the lack of consistent execution of his biweekly report. Josef then took time to review Douglas's progress versus some of his key objectives, pointing to opportunities for improvement. After some really healthy discussion, Josef set up a weekly check-in meeting with Douglas during which time the two of them would review progress versus restated expectations. Josef adjourned the meeting on a positive note.

You may be thinking at this point that Josef's behavior during the meeting was somewhat transactional in nature and that according to the presented literature on the topic, he should have maintained a transformational approach. To the contrary, recall our earlier discussion in chapter 3 regarding the most effective leaders. There we found that effective leaders must both provide what is to be done (transactional approach) and concurrently, offer vision and strategies (transformational approach) regarding how such may be accomplished.

Like Josef, Gerald was a double E. As such, the two of them had a great deal more in common than the remaining team members. However, being aware of this commonality, Josef took necessary precautions so as to avoid the potential perception of him showing Gerald favoritism. One of Gerald's key objectives was to work closely with the engineering department to assess the efficacy and efficiency of a recently installed automated weight management and filling system (WMFS). At the conclusion of this assessment, a coauthored report was to be compiled and issued to engineering, quality, and other department heads. Although the assessment approach was largely developed by Gerald, he relied heavily on his engineering peer (Brian) for details regarding system operations. The assessment lasted over several production shifts and identified several discrepancies relative to the original equipment manufacturers (OEM) stated capabilities and specifications. The findings were of particular interest to department heads, including the plant director, given that the cost for the WMFS continued to escalate following Summit Consumables's initial expression of purchase interest. Some believed that the purchase was too high despite the touted cost savings of system installation by the OEM. Needless to say, the report was used by some for political reasons—some viewing it as a step in the right direction and still others claiming that it was a total waste of money. As word got around about the report, so did misinterpretations about the actual findings. Aimed at getting everyone on the same page, the director of manufacturing requested that Josef's group schedule delivery of a comprehensive PowerPoint presentation discussing the assessment findings as well as any recommendations regarding next steps. Gerald seemed comfortable enough with preparing and delivering the presentation. Interestingly enough, Josef did not spend a lot of time reviewing the presentation, and Gerald did not engage anyone else in its preparation. The presentation was scheduled for Friday afternoon in the executive conference room in a building adjacent to the manufacturing building. Gerald had never visited this building conference room but felt confident that he knew where it was. The presentation was scheduled to begin at three o'clock and last for fifteen minutes.

On the day of the presentation, Gerald had trouble finding the conference room and arrived at five after three with his laptop. Upon arriving, he noticed, much to his surprise, that the room was filled to capacity. There was a lectern with a microphone, no cables to connect his computer, and a tremendous glare on the presentation screen. Josef jumped in to help with the cables and adjusted the shades to take care of the glare. After several minutes, Gerald's presentation began. After the fifteenth slide, the director of quality asked a question that seemed to catch Gerald by surprise—something well beyond his area of expertise that he had not contemplated previously. Gerald attempted to answer the question on the fly, and it seemed as though his response was a bit less than factual. Moreover, the level of detail that he provided was way too technical for the audience. Sweat began rolling from Gerald's brow, and his voice began to tremble a bit. At that moment, Gerald accidentally hit the forward button on his clicker/pointer, skipping to the end of the presentation. When he attempted to return to his last slide, he got a bit confused and could not find his prior stopping point. Josef, who felt really anxious for Gerald, suddenly looked away and to the floor. The audience became deafly silent. There wasn't a peep out of anyone. Finally, in a very perfunctory manner, Gerald said that he would send out a summary of the presentation to everyone and that he would be glad to respond to any questions via email. With that, Gerald turned off his computer, packed up, and left. The audience was stunned. Attempting to salvage the presentation, Josef was quite apologetic for the truly bummed presentation and stated that he would look for an opening on everyone's calendar in the coming days for a do-over.

I think it goes without saying that the weekend was long for Josef and even longer for Gerald. On Monday, however, Josef met with Gerald to talk about the presentation. Josef, realizing how visibly disappointed Gerald was, empathetically shared how he himself had not fared so well in the past in the area of presentation delivery. In addition to taking some ownership for not conducting a dry run of the presentation with Gerald, Josef pulled a brochure from his desk titled "The 5Ps for

Exhibiting Presentation Leadership." It so happened that there was a seminar being conducted on the coming Wednesday on the topic. Upon seeing this, Gerald asked if he could attend, and Josef responded, "By all means, I'll put a rush order on the purchase order."

At this point, you are no doubt thinking of the several missed opportunities (by Gerald and Josef) leading up to the presentation. Anticipating your thought process, I think you will find the next chapter very interesting.

CHAPTER 12

Introduction to the 5Ps for Exhibiting Presentation Leadership

THERE IS ONE very important aspect of communications that was not discussed earlier—formal presentations. And unfortunately, there is another stereotype associated with engineers suggesting that they do not present well. One key criticism in this area—as was the case with communications in the general sense—is that engineers often fall short in oral presentation when sharing a relatively technical topic, problem, or solution with a nontechnical audience. At the same time, giving an oral presentation can often mean the difference between getting the next promotion and being pigeonholed or overlooked in favor of someone else. Bolıvar-Cruz (2017) captured this sentiment by writing, "There is a strong consensus that future engineers, in addition to technical and mathematics skills, must also have skills that allow them to adapt to the new professional context, such as ... communicate, especially orally." Because of the criticality of this competency, this chapter in its entirety is dedicated to building necessary skills and competencies for the delivery of effective and efficient presentations. As you review this chapter, I encourage you to reflect on the incident with Gerald and consider what he could have done differently leading up to and during his presentation.

In my book *All the Way to the Top: A Practical Guide for Corporate and Business Leadership*, I discuss at length the 5Ps for exhibiting presentation leadership, including a mock application of these very powerful presentation tools under the guise of key characters employed in a fictitious consumer products corporation—Summit Consumables. Given the very crucial role that presentations play in making or breaking careers, I have included excerpts from that writing as follows: "Many people—even the most accomplished public speakers feel everything from slight nervousness to outright terror before they give a presentation ... one study revealed that an alarmingly high number of people would rather have unnecessary surgery than give a speech" (*Harvard Business Review*). This discussion is not intended to coach you in developing cool PowerPoint slides. Many resources and presentation templates exist to assist in that area. It is instead aimed at addressing a much more fundamental yet essential component of presenting effectively—the exhibition of leadership ahead of and during the formal presentation.

We will begin the discussion of presentations after paying our customary homage to *Merriam-Webster (2007)* in the areas of oration, speech, and presentation. "What do oration and speech have to do with delivering a PowerPoint presentation?" you ask. Well, simply put, everything! To this end, *oration* is defined as "an elaborate speech" (*Merriam-Webster'* 2007). Speech is defined as "the act of speaking ... talk, conversation ... the power of expressing or communicating thoughts by speaking" (*Merriam-Webster* 2007). Now for our final definition, presentation is defined as "the act of presenting ... the act power or privilege ... symbol or image that represents something ... a descriptive or persuasive account ... as by a salesman of a product" (*Merriam-Webster* 2007). We will return to these definitions at the end of this section. For now, I'd like to set context through the sharing of excerpts from some of the most famous speeches of all times.

The first of these was given by Abraham Lincoln during his first presidential inaugural address on March 4, 1861.

> Fellow-citizens of the United States: In compliance with a custom as old as the government itself, I appear before you to address you briefly, and to take in your presence the oath prescribed by the Constitution of the

United States to be taken by the President before he enters on the execution of his office ... I do not consider it necessary at present for me to discuss those matters of administration about which there is no special anxiety or excitement ... apprehension seems to exist among the people of the Southern States that by the accession of a Republican administration their property and their peace and personal security are to be endangered. (Abraham Lincoln, www.loc.gov)

Speaking to the United States following his election and amidst the Depression, Franklin Roosevelt, during his March 4, 1933, inaugural address said,

> This is a day of national consecration ... and I am certain that on this day my fellow Americans expect that on my induction into the Presidency I will address them with a candor and a decision which the present situation of our people impels. This is preeminently the time to speak the truth, the whole truth, frankly and boldly. Nor need we shrink from honestly facing conditions in our country today. This great Nation will endure as it has endured, will revive and will prosper. So, first of all, let me assert my firm belief that the only thing we have to fear is fear itself. (Franklin D. Roosevelt, www.millercenter.org)

Mahatma Gandhi, as an advocate for Indus freedom from British rule beginning in 1915, said, "Let me place before you one or two things ... I want you to understand two things very clearly ... I ask you to consider it from my point of view, because if you approve of it, you will be enjoined to carry out all I say ... let me assure you that I am the same ... as I was in 1920 ... I attach the same importance to nonviolence" (Mahatma Gandhi, Daley 2013).

Susan B. Anthony, speaking regarding women's suffrage on January 23, 1880, said,

> Friends and fellow citizens ... I stand before you under indictment for the alleged crime of having voted at the last presidential election, without having a lawful right to vote ... it shall be my work this evening to prove to you that in thus doing so, I not only committed no crime, but instead simply exercised my citizen's right, guaranteed to me and all United States citizens by the National Constitution beyond the power of any State to deny. (Susan B. Anthony, Daley 2013)

And finally, in his address to the nation on the war in Vietnam on November 3, 1969, Richard M. Nixon said,

> Tonight I want to talk to you on a subject of deep concern to all Americans and to many people in all parts of the world—the war in Vietnam ... I believe that one of the reasons for the deep division about Vietnam is that many Americans have lost confidence in what their Government has told them about our policy. The American people cannot and should not be asked to support a policy which involves the overriding issues of war and peace unless they know the truth about that policy. (Richard M. Nixon, www.presidency.ucsb.edu)

Upon reviewing these excerpts, one thing becomes quite apparent. Each of the speakers seeks to achieve an end. Abraham Lincoln sought to assuage concerns regarding the potential loss of the most fundamental human needs—to maintain their families, homes, properties, and lives. Franklin Roosevelt sought to transform the nation's view of its current situation of depression (after the stock market crash) to one of prosperity and renewal. Mahatma Gandhi sought to reassure the India Congress Committee of his unwavering belief in nonviolence as a means through which change may be effected. Susan B. Anthony sought to defend herself against allegations of crime resulting from her voting. And finally, expressing policy concerns, Richard Nixon sought to reconcile division in the nation regarding the Vietnam War.

While there was much more to be conveyed within each of the speeches, the fundamental purpose of each speech was immediately clear. Thus, *purpose*, the first of our 5Ps, is defined as "something set up as an object or an end to be attained" (*Merriam-Webster* 2007). You may recall being present in a meeting or other environment during which the speaker seemed to ramble on until you had to wonder where his speech was headed. Usually, this means the speaker has not made the purpose of the speech clear.

In a corporate or business setting, the purpose for a speech or presentation is often made clear by the requester of such. Purpose, used in the context of the 5Ps, is all about defining what you are presenting and why. Are you seeking approval, disseminating goals and objectives to your team, advocating a position or recommendation, or simply providing an update? As evidenced by the excerpts provided, whether you are being asked to provide a solution to a critical business issue or simply provide a status update, the purpose should be clear and evident at the outset of the presentation.

Having defined the purpose, you now must implement the necessary steps—mechanical, qualitative, and quantitative—for fleshing out your presentation through *preparation*, the second of our 5Ps. General Colin Powell stated, "There are no secrets to success. It is the result of preparation, hard work, and learning from failure." According to *Merriam-Webster (2007)*, *prepare* is defined as "to make or get ready … to get ready beforehand … to put together … to put into written form" (2007). Good preparation necessitates that each of the following questions and points be addressed.

Who are the participants expected to attend the presentation? Knowing who they are will allow you to modify your delivery to fit the audience. As previously mentioned, you've no doubt heard the adage "When in Rome, do as the Romans do." Well, as the presenter and speaker on center stage, there is little that can be more damaging to your presentation than speaking around, over, or above your audience. This is sometimes the case when a highly technical individual is asked to present to a nontechnical audience.

What perspectives might the key participants and stakeholders bring to the presentation that may be in conflict with those of your own? Similarly, if you are but one of several presenters to the same audience, it is also important to be aware of the topics that will be covered by your peer presenters. In the event you are presenting the same topic, the worth of your delivery may result from a comparison with the peer who addresses the same topic. So if flexibility exists regarding the purpose of your presentation, being cognizant of the fact that this potential may afford you the opportunity to either modify your presentation or collaborate with the other presenter(s).

In preparing for any presentation, it is always important to be aware of the water cooler conversation as it pertains to your particular topic. In short, what are the key stakeholders saying about the issue you will be discussing?

Time and circumstances permitting, once you are able to identify dissenting perspectives and opinions, you can then visit those stakeholders to not only better understand their positions but also determine how rigid they might be in those positions and possibly shift their thinking such that it is more compatible with yours.

Alternatively, if applicable, modify your position to one more aligned with the expectations of your audience. Reaching out to your audience ahead of the presentation is a critical step because the last thing you want to occur is for your presentation to be derailed by new information surfacing, particularly if it might serve to undermine one or more of your key arguments.

Knowing and understanding the perspectives of the audience has been a strategy executed by many great leaders. Indeed, during the preparation of Lincoln's inaugural address, he recognized that discussions of "matters of the administration" took a back seat to concerns of "property, peace and personal security." Likewise, Roosevelt recognized during the preparation of his inaugural address that the nation had grave concerns or "fears" when considering its then current state.

What is the time line for the presentation, and how much time is allocated to you? On occasions, I have attended a presentation where the presenter's slides contained one or more errors (typically misspellings) and some other errors. My personal experience suggests that when a presentation is pulled together at the last minute, the risk for typographical errors or erroneous data being included is increased almost exponentially. As the presenter, you should always know the date of the presentation and should allow enough time in your preparation time line for that date to be compressed. Despite our good intentions, the typical brain functions best when it's not under the gun to deliver. Likewise, if your presentation is scheduled for ten minutes, don't design a presentation that requires exactly ten minutes for you to completely cover all your slides. Invariably, you will either receive a question or unexpectedly ad-lib a bit, resulting in your presentation running over the allotted time.

What is the story that you want to tell? Reynolds (2008) said of storytelling, "Good stories have interesting, clear beginnings; provocative, engaging content in the middle; and a clear conclusion." Now that you understand the purpose of your presentation and have identified the attendees as well as their perspectives on the topic, it is time to pull it all together. While it may seem a bit redundant, even though the purpose of your presentation was given to you, it is still a great idea if not a requisite action to restate it up front in your presentation. The intent of this discussion is not to show you how to design presentation slides. But early on in your presentation, certainly not later than a slide in, you should restate the purpose. Don't include every word that you plan to share during the voice-over (your actual speaking) as this will make for a busy slide. But do provide enough information so that those who prefer to read your purpose slide have a good understanding of the presentation purpose.

Once your purpose has been clearly defined, the next step is to share a bit of background about the issue to be addressed. For example, if the purpose of your presentation is to resolve issues with floundering sales in a certain region, you might want to show historical data leading up to the current situation. In doing so, you would be sure to point to any interesting or assignable causes that you might be able to make use of for turnaround later in your presentation.

All the while, you are walking your audience through the issues and leading them to your next focus area. In this case, you might consider discussing alternative courses of action that have the potential to resolve the depressed sales issue. But you want to do so succinctly, addressing no more than three total options and the pros and cons associated with each. (Often the pros and cons discussion includes reference to a cost-benefit analysis or other quantitative or qualitative means for alternative differentiation.)

During a presentation there is often insufficient time to discuss a lot of information. Nor is there sufficient time to discuss a little information in great detail. Consequently, the options should be reviewed at the right level of detail based on your execution of the strategy previously discussed (knowing your audience and their perspectives).

Next you advance your recommended course of action. Since this represents the pinnacle of your presentation, it is extremely important that you think it through so that it will be seen as the no-brainer option that will withstand the pressures of critique, scrutiny, and questions posed by the audience.

So you're done, right? Not so fast! Although you have shared the issue and its respective solution, you have not yet discussed the implementation plan. This is one of the key points that speakers often overlook—solution implementation feasibility.

While a comprehensive review of a feasibility study is beyond the scope of this book, a few critical points are noteworthy. In particular, depending on the type of project or alternative, aspects of its feasibility might include the following: technical, economic, schedule, and operational. As the topic headings imply, the technical feasibility study has to do with technology and risks. Economic feasibility has to do with resources and respective constraints. Schedule feasibility has to do with the time required to execute the recommendation, and lastly, operational feasibility concerns problems or issues (if any) associated with the alternative such as management resistance (cs.toronto.edu).

This is not at all to suggest that the presenter provide a comprehensive feasibility analysis report during the presentation. It is, however, incumbent upon the presenter to share enough information to achieve comfort among the audience that the recommended path forward has been fully vetted.

At this point, you have completed the fundamental requirements of story development. Some presenters include as the final slide a Q and A prompt. While this is a good practice, my experience suggests that rarely will the audience allow you to reach the end of the presentation before asking questions or seeking clarification.

Know the role of your slides. During your presentation you should by no means simply read your slides. Your audience could do that by receiving a copy via email. Instead you should allow your slides to guide your conversation with the audience. For example, if on one of your slides you wanted to display the difference in NPVs for two project alternatives, you might simply show the NPV for each project. But during your discussion, you might want to review the cash flow sequence or the fact that you also considered IRR and PI as well as payback. In this way, the attention is on you, the presenter, and not your slides. (By the way, I know it sounds like a broken record, but I cover in detail NPV, IRR, and other financial terms in my book *All the Way to the Top: A Practical Guide for Corporate and Business Leadership*.)

So how do you commit to memory all the information that you want to share without putting it on a slide? Great question! Something that has worked well for me is the *notes* feature in PowerPoint to capture what is to be said, and I also make mental connections between the slides and the words in the notes section. Typing what you are going to say in the notes section as closely as possible to how you plan to communicate it during the actual presentation will go a long way toward eliminating the perfunctory delivery of attempting to recall and convey data or numbers in a manner inconsistent with your typical communications style. Repetition is a great way to solidify the connection between notes pages and actual slides.

Test-drive your story over and over again! Many guinea pigs (people to listen to your dry run) are available, and you can rely on them for unbiased feedback. An administrative assistant is always a good choice for this approach along with peers and family members. Select a mix of those who know absolutely nothing about your topic and those closest to and most knowledgeable about it.

The value of this approach is twofold. First, by selecting someone who knows nothing about your topic, you can assess whether or not your story is logical and generally cogent. Despite the rhetorical and typical nature of inaugural addresses, there have been instances where there were clear points that allowed the audience to get it despite having little to no firsthand knowledge of the subject matter. One example is Roosevelt's "the only thing we have to fear is fear itself" comment. Likewise, regarding Kennedy's January 20, 1961, inaugural address, consider his words, "My fellow Americans: ask not what your country can do for you—ask what you can do for your country" (Daley 2013). I submit that very few listeners or readers failed to comprehend the gravity of those spoken words.

Second, selecting someone who is very close to your topic serves as a sort of acid test and fact check. Despite our best intentions, we are all fallible human beings, and as such, we may not get it right all the time. Telling your story to someone who has been there and done that often reveals holes in your story that you may not have recognized or anticipated.

"Who cares?" you say. "No one in the audience will know as much as I do about my topic." Don't kid yourself! Typically, particularly in possibly contentious situations, those attending a presentation may ask a person with whom they associate or more often someone who works for them to join the attendee in the presentation to sort of keep the conversation honest. Or sometimes an attendee may review the key topic areas of your presentation before attending in hopes of becoming an *overnight expert*. So as the presenter, you should never go into a presentation thinking that you can simply gloss over a topic without anyone noticing.

Know your knowledge limitations. Don't become an overnight expert. Similar to the second point discussed in item 6, often in an effort to impress or otherwise fly solo, presenters attempt to incorporate information in a presentation about which they have only superficial knowledge. There are two issues with this. First, while it may seem easy to have a conversation with an expert on a topic and believe you understand its ins and outs, recalling that information in the heat of the battle (meaning during a live presentation) is something different. For example, if you were told that a certain process change such as reducing the variation of an input part actually improved the quality of a product, which should result in fewer customer complaints, this may seem simple enough to regurgitate during a presentation. But what if someone in the audience asked you to talk about the statistical confidence level that assertion was based on or to discuss whether the statistical inference was predicated on parametric or nonparametric statistics? How would you respond? Better approaches would be to avoid this comment altogether or to build into your presentation enough time for an expert to offer a slide or two on the topic. If the latter approach is taken, remember that it is your presentation, and you are working with a defined time slot. Thus, some coaching of the additional participant may be in order.

A slight spin on the foregoing comments is to recognize that invariably and despite your most diligent efforts, questions will be asked and comments offered to which you may not have the correct answers. When that happens, just respond by saying, "Good question. I haven't looked at that but will certainly consider it." Or you may say, "Perhaps we can talk more about that following my presentation."

These and similar responses work well for questions that are not germane to the logic used in your presentation. As long as the question or comment doesn't significantly weaken your proposed position, the audience should be accepting of your use of one of these or similar responses. However, use them in a measured and calculated manner.

I once attended a conference where several participants holding PhDs offered presentations. I recall two particular presenters. The first of these was someone who studied computer science. The other was in the field of business management and strategy.

During the first presentation, in reply to almost every question, the presenter responded, "That question is beyond the scope of my research," even giving that reply to questions that appeared to be germane to the key points discussed.

During the second instance, I asked a question about psychometric modeling. I believed that if the presenter had considered this model, it may have significantly changed the presented perspective. In response to my question, the presenter said, "Here's my card. Give me a call, and perhaps we could discuss a collaboration over a cup of coffee."

While I'm not exactly sure what the presenter meant, I was pretty sure that the audience recognized a potential hole in his presentation exposed by my question, which was not at all intended to undermine what was otherwise a great presentation at all. (I never made use of the card he gave me.)

So diligently prepare, anticipate some tough questions, use certain responses in a meaningful way, and know when to play the "Let me get back to you" card.

Location, location, location. Where will the presentation be delivered? Several years ago I was asked to introduce a revered Harvard professor at a location with a seating capacity for more than thirty-five hundred attendees (speaking to a crowd this large would be a first for me). Initially, the thought of delivering this introduction sort of made my stomach feel a little queasy. However, that queasy feeling was short-lived as one of the first things that I did was to arrange a practice session at the facility. Visiting it afforded me the opportunity to test out the sound system, begin to create eye focus with certain sections of the theater seats (which I linked to certain points in my introductory speech), observe the lighting, and become familiar with the location of the lectern relative to the curtain from which I would emerge. Most importantly, it gave me comfort that on the evening of the event, I would be introducing a renowned Harvard professor in a facility that I had not only become acquainted with but in which I had also repeatedly practiced.

My introduction was comprehensive (and I admit it was a bit long), but with the practice under my belt, I was feeling pretty good. Following the real-time introduction, the professor gave me a big hug, and the applause was tremendous. To top it off, the next morning I received in my in-box a copy of an email addressed to my then boss's boss raving about my introduction and also praising my "excellent representation of the company." That the sender of the email held a senior executive-level position in a field where public-speaking ability served as a make-or-break core competency was simply icing on the cake when considering the complimentary comments.

I attribute at least 80 percent of the success of that delivery to having visited and practiced in the facility ahead of the event and to getting others to listen to and critique my story before the event. Much of the remaining 20 percent may be ascribed to my speech writer.

Although the discussion of location in this section referred specifically to a speech delivery, I followed a similar strategy for presentations as well, particularly when the location was one with which I was unfamiliar.

These steps may not occur in the sequential order listed. For example, you may learn of the location for the presentation immediately when it is requested. The key point, however, is that you use these steps as predecessors to the remaining three of the 5Ps.

Think about the following references and scenarios to see if you can determine the common theme. Some of them may sound dated. But I believe them to be germane to the core point. Consider the following:

- Kenny Rogers's song "The Gambler,"
- the movie *Jerry Maguire*, where Dorothy Boyd said, "You had me at hello,"
- a salesperson who says to you after a locked-in deal on a new car, "I would have reduced the price another few hundred dollars if you had just held out for a little while longer," and
- a coworker who says, "Wow, looks like the boss is in a pretty foul mood today."

The answer brings us to the third of the 5Ps, which is *perception*. All four scenarios had to do with someone failing to observe a verbal or nonverbal cue that, if caught, may have influenced someone's next action.

In the Kenny Rogers song, it has to do with a gambler reading the faces of poker players and those players doing likewise.

In the Tom Cruise movie *Jerry Maguire*, Tom's character (Jerry) walks into the house where Renee Zellweger's character (Dorothy) is present, and he says hello and proceeds to talk about how he really cares for her and wants his wife back. As he continues to talk, Dorothy interrupts him, saying, "Shut up … you had me at hello."

The salesperson, if pushed for a greater discount, would have caved by several hundred dollars. And the boss, whether verbally or nonverbally, may have exhibited a petulant disposition.

Merriam-Webster (2007) defines *perception* as "an act or result of perceiving … awareness of one's environment through physical sensation … ability to understand." During a presentation there will be many nonverbal cues and possibly some verbal ones that the presenter should recognize. Dubrin (2010) says of nonverbal and verbal communications, "Effective leaders are masterful nonverbal as well as verbal communicators." Your leadership responsibility should not diminish while presenting. To the contrary, there is perhaps no greater visible opportunity in which to display it.

The presenter's nonverbal communications are just as significant. If, for example, the presenter responds uncertainly to a question, the asker and possibly other attendees may observe the presenter's manner of responding, resulting in a lack of speaker credibility.

Dubrin (2010) also says, "A self-confident leader not only speaks and writes with assurance but also projects confidence through body position, gestures, and manner of speech." Dubrin points to several observable attributes of a confident leader (speaker), including the following:

- standing with toes pointing outward rather than inward;
- using an erect posture when walking, standing, or sitting;
- speaking at a moderate pace with a loud, confident tone;
- smiling frequently in a relaxed, natural-appearing manner;
- maintaining eye contact with those (in the audience); and

- gesturing in a relaxed, nonmechanical way, including pointing toward others in a way that welcomes rather than accuses.

It follows that despite all the preparation to nail the purpose and adequately prepare, insufficient maturity in perception or perceptiveness can make or break an otherwise great presentation. But even after you have defined the purpose, adequately prepared your presentation, and learned all there is about perception, what then? Have you completed all that is necessary to deliver a stellar presentation?

The answer is not quite. Many times despite your best efforts at preparation, there remain those who see the world a bit differently. And unless those attendees buy into your views and perspectives, your job may not be complete. So you as the presenter have an accountability to persuade others to see things just as you do, which represents the fourth of our 5Ps of presenting—*persuasion*.

From our early years when we induced our parents to see things our way by not grounding us or buying us that special toy despite our undeserving behavior, we were learning and fine-tuning our persuasiveness. According to *Merriam-Webster* (2007), that means "to win over to a belief or course of action by argument or entreaty." Here *argument* does not refer to the emotional discussions that may result from disagreement.

To better understand the distinction, let us consider a philosophical approach to argumentation. An argument consists of premises followed by a conclusion. These premises may be viewed as statements leading up to a claim of some sort. The premises are often referred to as propositions. The claim may also be termed the conclusion, and it is typically preceded by words such as *therefore* and *consequently*.

Consider the following example:

- It is raining outside.
- Rain is wet.
- Therefore, the ground is wet.

In this simple argument, "It is raining outside" and "Rain is wet" act as the premises, and "Therefore, the ground is wet" serves as the conclusion.

Statements may also have truth value, which tests whether or not the statement is true based on what we know and understand to be correct about the world in which we live. A statement asserting that the sun is green, while it is indeed a statement and could also be predicated on one or more premises, has very little truth value because we can see that the sun is not green.

Arguments may be deductive or inductive. Their validity is linked to the premises used. An example of a deductive argument is as follows:

- All human beings need water for survival.
- You are a human being.
- Therefore, you need water to survive.

This argument is valid because the premises lead to the conclusion. It is also sound because according to our worldviews, the argument (including the premises and conclusion) is also true. Validity and truthfulness are not synonymous. Whereas validity in this context refers to whether or not the argument structure is correct, truthfulness refers to the correctness or incorrectness of the argument itself.

In contrast, an inductive argument seeks to gain buy-in that the conclusion most likely follows from the premises. These arguments cannot be termed valid or invalid as was the case with deductive arguments. An example of an inductive argument would be as follows:

- Some school-age children born in the South play two different musical instruments.
- Consequently, school-age children born in the South play two different musical instruments.

This argument generalizes that school-age children in the South play two different instruments, which may or may not be true when considering all school-age children born in the South (Copi 2002; cse.buffolo.edu).

Other key terms associated with arguments are not discussed here, such as inductive argument cogency, a further analysis of deductive arguments, and the premises included therein. Further inquiry in this area is encouraged. However, the key takeaway from this philosophical discussion is that the art of persuasion is predicated to a large degree on the presenter's ability to offer sound and logical arguments and also to assess arguments offered by others for their validity and truthfulness.

On June 12, 1987, Ronald Reagan, using logic and sound reasoning, exhibited one of the greatest acts of persuasion ever documented when he challenged General Secretary Gorbachev to dismantle the Berlin Wall. Here is a key excerpt from the speech:

> There is one sign that the Soviets can make that would be unmistakable, that would advance dramatically the cause of freedom and peace ... General Secretary Gorbachev, if you seek peace, if you seek prosperity for the Soviet Union and Eastern Europe, if you seek liberalization ... come here to this gate ... Mr. Gorbachev, open this gate ... Mr. Gorbachev, tear down this wall (Daley 2013).

At the time, many believed—and some continue to believe today—that the dismantling of the Berlin Wall shortly after Reagan's request exhibited not simply the power of persuasion but also the power of the United States of America.

Interestingly, many dictionaries list the word *influence* as having a close relationship to persuade. Ivancevich and Matteson (1993) also noted a relationship between influence and power, stating that "power is the potential to influence, while influence is power in action."

Therefore, the adept presenter will recognize the opportunity to demonstrate power—leveraging persuasion and influence—during the presentation. Its execution represents the fifth of the 5Ps—*power in action*.

Unlike our prior discussion in this area, *power*, as it used here, is not an individual attribute. It is instead predicated on interpersonal interactions.

With every presentation comes the opportunity to interact with others, which also allows for the demonstration of interpersonal skills. As we now know, power may be categorized as legitimate, reward, coercive, expert, and referent. And while the first three may be viewed on an organizational level, our focus here is on expert and referent power (also known as personal power).

As previously discussed, expert power, as the name implies, suggests that the leader has special expertise in a particular field. Referent power suggests that the leader possesses charismatic personality traits. A presenter exhibiting both of those kinds of power is more likely to achieve power in action during a speech or presentation (Ivancevich and Matteson 1993; Dubrin 2010; Bateman and Zeithaml 1993).

To further solidify the concept of power in action, consider some comments that I heard while listening several years ago to an audio program on verbal skills building. "When Cicero spoke, people marveled … when Caesar spoke, people marched" (Cato, Roman Politician, Elster 1995). This is a clear example of power in action. This is not at all to suggest that you as a boss should convey the simplest direction as though you were sending troops into battle or that you as a subordinate should respond to direction or questions with an eloquent speech that you spent all night compiling and committing to memory. It does mean, however, that you communicate clearly, decisively, and with conviction.

Staying true to my belief in Ernest Hemingway's aphorism to show people everything but tell them nothing, figure 12–1 reflects the relationship between and among the 5Ps.

Figure 12–1. Interdependence of the 5Ps

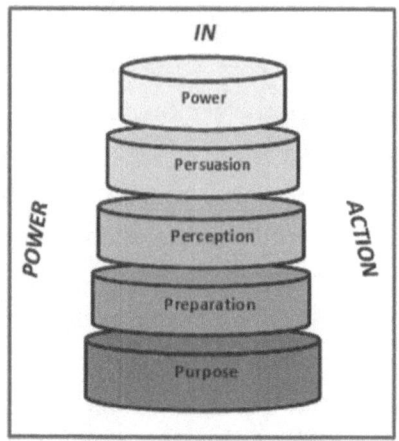

Viewing the diagram from bottom to top, note that purpose is the first stage followed by preparation, perception, and persuasion, ending with power. This diagram could also be viewed as analogous to a powerful rocket inasmuch as each stage facilitates implementation of and shares interdependence with the next stage's function and implementation. Just as the first stage of a rocket is required to generate the thrust for liftoff, so the purpose of a presentation is required in order to have any chance of success. Likewise, the middle stages of our 5P rocket must be fully completed and operational, thus enabling a successful mission (presentation). The fifth and final stage may be viewed as the command module through which the presentation mission (execution of power in action) is ultimately accomplished. Earlier in the chapter, we were given certain dimensions associated with effective communications based on a West Point training approach. Among those were *organization* and *substance*, both of which are integral to the diagram and its practical application.

It is my hope that the linkage between oration, speech, and presentation (as was introduced at the outset of this chapter) is now quite clear. The relationship between oration and speech should be obvious. And persuasion, perception, and power were all used to define speech and presentation. Thus, through their definition, we have accounted for three of the 5Ps discussed previously. The remaining two (purpose and preparation) are assumed to be very fundamental enablers for delivering effective presentations.

If the linkage (oration to speech to presentation) did not resonate with you, the key takeaway from this discussion about power in action is that as the presenter, you are standing center stage among your audience, and you are the one who will be perceived as having the answers. It's simply a matter of making use of your expert and referent power to demonstrate power in action as you move the audience to where you know they should be as reflected by the story that you tell and the conviction with which you tell it.

It's now time to test your understanding of the concepts previously discussed.

Example 1

Consider the discussion regarding the 80/20 rule. Review the following excerpts, which were taken from an article titled "Legalese Just Gets in the way of Effective Communication." Rewrite the excerpts as concisely and efficiently as possible, being careful not to lose the original meaning (McElhaney 2010). Good luck!

1. Would you indicate please, for the benefit of the judge and jury, what you did in response to the petitioner's request that you ameliorate the conditions created by your company's failure to make timely delivery of the automatic current interrupters?
2. Directing your attention to the 23rd of May, did you have occasion to discuss the terms of the agreement that would apply in the event of noncompliance on the part of the petitioner in this cause of action concerning its obligations to the respondents?

Believe it or not, both of these excerpts were reported as being actually shared during courtroom proceedings.

Example 2

Consider the following simple argument (*Introduction to Logic*, Harry J. Gensler):

- If you overslept, you will be late for work.
- You didn't oversleep.

Referring to the discussion regarding argumentation, which of the following represents an accurate conclusion based on the two premises given?

- You're late.
- You aren't late.
- You didn't oversleep.
- None of the above.

Example 3

You have been asked to provide a presentation to your boss and her peers to talk about a critical project that you have been working on for several weeks now. You have run into a slight problem, but you have also identified a solution that requires the support of your boss and a key peer department. You were also informed that the presentation would take place in the staff conference room, which you had never before visited. The request came to you via email. At the end of the email, your boss reiterated his support for you and said that this would be a great opportunity for her peers to get a good look at you. With this in mind, please complete each of the following steps:

- List two or three steps that you would take in developing the purpose for your presentation.
- List several questions/issues that require addressing as you implement the preparation stage for your presentation.
- Assume that you have developed your presentation.

- What role might perception play during your presentation delivery?
- Anticipating your presentation, what techniques might you employ in the area of persuasion?
- What expectations will you have of your audience regarding your call to march, and how will you know that you exhibited power in action?

Please feel free to contact me at leadershiplmc.com should you have questions or comments regarding these examples.

CHAPTER 13

Bringing It All Together

I REALIZE THAT WE have covered a lot of ground. Many years ago while pursuing my MBA, I had a professor who, after reviewing lots of information in an advanced financial management course, said, "Okay, when we're done, I'm going to bring it all together." I must say that in retrospect, he was exactly right. So stealing his line, I'm going to attempt to bring all the information that you have reviewed together in just a few more pages.

In chapter 1, we saw how more of the top-performing CEOs possessed engineering degrees than MBA degrees. Incorporating scholarly review and perspective on the topic, we also discussed the necessity for US businesses to have more engineers with knowledge in management and leadership. Employing a very simple electronic logic circuit, we were able to contrast at the most fundamental level the stark differences between binary approaches to problem identification and resolution and not so much black-and-white approaches (that which we termed the darker shades of gray).

In chapter 2, you were introduced to scholarly characterizations of leadership in the broad sense with peppered-in views of engineering leadership. In doing so, you noticed how some authors choose to circumscribe the leadership sphere of engineers—an approach that I strongly oppose. You were also exposed to leadership systematization (inputs, processing, and outputs) before diving into discussions about theoretical leadership models, including contingent leadership, path-goal theory, the Vroom-Jago model, and leader-member exchange. We then discussed what was appropriately termed leadership levers—teams, feedback and coaching, and how the leader made use of each of these to achieve optimum leadership effectiveness.

Discussions then shifted to emotional intelligence and how maturity in this area enabled further leadership success in decision making and facilitating resonance. In particular, we discussed the four domains in detail—self-awareness, self-management, social awareness, and relationship management. We also talked about the need for leaders to self-reflect and replenish the well. You were then introduced to full-range leadership theory, focusing on transformational and transactional styles. Employing Maslow's needs hierarchy, you were exposed to the role that psychology plays in leadership and the associated cognitive process through employing the performance appraisal example. Several leadership assessment models were then discussed, including the five-factor model, three-factor model, and the Myers Briggs Type Indicator.

In chapter 3, we revealed the role that power plays in influencing others and discussed the five power bases (legitimate, coercive, reward, expert, and referent) and how each may be reflected in leadership behaviors. Linking power to leadership styles, we then made the connection between transactional leadership and three of the power bases (legitimate, coercive, and reward) and then linked the remaining two bases (expert and referent) to transformational leadership. We also made the case, graphically and otherwise in the context of transformational and transactional styles, for the leader's need to exhibit all power bases. Chapters 4 through 6 raised critical research questions regarding the role—if any—of engineering education on perceived leadership style and the respective hypotheses for addressing these questions. We also set the stage for the research (e.g., parameters, methodology, procedures employed, etc.), and you were introduced to full-range leadership theory in greater detail.

Presented in chapter 7 were all aspects of research execution. A review of population demographic information was presented, and the testing groups were further defined. Data analysis, which employed parametric statistical approaches, was reviewed with great granularity when considering transformational and transactional leadership style constituents in route to addressing the potential influence of engineering education on perceived leadership style both across tested groups and within such groups. Reference was also made to nonparametric statistical approaches presented in appendix G to buttress parametric testing.

In addition to summarizing the prior chapter research conclusions and introducing follow-up testing regarding a key transactional leadership constituent, Chapter 8 addressed the relevance and generalizability of the research results as well as the expressed limitations. It also advanced areas for potential future research, such as the role that gender or experience might play in perceived leadership styles.

Chapter 9 introduced the hypothetical corporation and a particular character named Josef Brilliant, an electrical engineer with a degree. Josef's character was used to emphasize the previously discussed survey-reported development opportunities for engineers (e.g., effective communications) as were identified in his performance review. You were also introduced to a key 5CLIM component (cognizance) and the very important role that such plays in managing priorities on a personal or departmental level. We saw how Josef was able to leverage his expert power to resolve manufacturing issues and how doing so positioned him for future leadership opportunities.

In chapter 10, we introduced another hypothetical scenario involving Josef, who was now operating in an organizational leadership position, and revealed the results of a 360-degree assessment that mirrored the previously discussed research results. You were exposed to both quantitative and qualitative aspects of the assessment. You were also once again introduced to the IPO model. This time, however, you were able to see how the IPO model could be used for process improvement both on a micro and macro basis. You were also able to make the connection between the customer used in the commercial product/goods sense and the same used in the context of intradepartmental customer/supplier relationships.

In chapter 11, we saw how Josef conducted himself (leading from behind) relative to his seven direct reports in response to received (quantitative and qualitative) feedback. Here, practical examples were employed to make the case for the value added by focusing on individualized consideration on a contingent basis. And in doing so,

Josef employed key transformational and transactional constituent actions.

The final employee scenario presented exposed another documented development opportunity (stereotype) for engineers in the area of oral presentations. This then provided fodder for chapter 12, where you were introduced to the 5Ps for exhibiting presentation leadership. Thus, you were exposed to purpose, preparation, perception, persuasion, and power and how each, all of which are operating on an interdependent basis, may be employed to yield effective and efficient presentations.

As a point of interest, you should be aware that the degree to which Josef seeks to develop employees is contingent on the employee as well as the employee's organization level. For example, you would not expect an employee operating at the manager level and reporting to a director to exhibit the lack of planning as was the case with Gerald. Thus, in general, the leader's style should become less directive and participative as he or she moves up the organization ladder.

While the focus of this book was engineers as leaders, it should be obvious that—as is the case with much in life—not all engineers possess the necessary passion to succeed in leadership, some of which may be driven by the engineer's worldviews. Indeed, according to Brazil and Farr (2009), "Equally important is the understanding that each leader brings to the situation a unique level of prior development attained by genetics, childhood upbringing, and adult experiences."

There is no doubt that engineers add tremendous value operating solely within their discipline. Weingardt (1994) espoused this view as well, writing, "Engineering is a noble calling. We create wealth and help maintain the standard of living for everyone. We turn ideas into reality: a design into a product like a building, a bridge, a car or a spaceship, or a computer chip into a machine that does something." Unfortunately, this view also reinforces the engineering stereotype, namely the person with the thick glasses working in a cubicle adorned with blueprints, prototype models, and other technological gadgets. Consistent with the foregoing discussion and research results regarding the potential influence of engineering education on leadership style, I believe that engineers offer leadership potential not circumscribed by stereotypical technical competencies. Weingardt (1994) ostensibly shared this view in writing,

> We desperately need to change this state of affairs. We need to move up the "economic food chain." Not only must we offer the country and the world our expertise as engineers in making things run but we must also become much more involved in running things. Our involvement can and will, ultimately, benefit us as individuals and our profession at large, but we also have a great deal to offer our communities and our world … to accomplish this, we need engineers in leadership in society."

At this point, you may be asking how to gain additional exposure to leadership applications on a more practical and business-oriented level. I would be remiss if I didn't suggest that you read my previously referenced title *All the Way to the Top: A Practical Guide for Business and Corporate Leadership*, which may be found in bookstores or on Amazon or BarnesandNoble.com. It draws on key lessons learned during my pursuit of my MBA, and it includes many more real-world experiences and scenarios. I conclude with the following comments. There is an adage that says, "You can lead a horse to water, but you cannot make him drink." You have indeed been led to the water. It is now up to you to do more than simply lead. I challenge you to truly change the world!

REFERENCES

Adams, K., and J. Keshwani. "Cross-Disciplinary Service-Learning to Enhance Engineering Identity and Improve Communication Skills." *International Journal for Service Learning in Engineering, Humanitarian Engineering and Social Entrepreneurship* 12, no. 1 (2017): 41–61.

Allison, S., Bastiampillai, T., Goodall, A., and Nance, M. "Expert Leadership: Doctors Versus Managers for the Executive Leadership of Australian Mental Health." *Australian and New Zealand Journal of Psychiatry* 49, no. 5 (2015): 409–411.

Anantatmula, V. "Project Manager Leadership Role in Improving Project Performance." *Engineering Management Journal* 22, no. 1 (2004): 13–22.

Antonikas, J., B. Avolio, and N. Savisubrananiam. "Context and Leadership: An examination of the nine-factor full range leadership theory using the multi-factor leadership questionnaire." *The Leadership Quarterly* 14 (2003): 261–95.

Armstrong, A., and M. Nuttawuth. "Evaluating the Structural Validity of the Multifactor Leadership Questionnaire (MLQ), Capturing the Leadership Factors of Transformational-Transactional Leadership. *Contemporary Management Research* 4, no. 1 (2008): 3–14.

Atwater, L., B. Avolio, and B. Bass. "The Transformational and transactional leadership of men and women." *Applied Psychology: An International Review* 45, no. 1 (1996): 5–34.

Aguayo, R. *Dr. Deming: The American who Taught the Japanese about Quality.* New York: Fireside, 1990.

Armstrong, M. *Armstrong's Handbook of Performance Management—An Evidence-based Guide to Delivering High Performance: Fourth Edition.* London and Philadelphia: Kogan, 2009.

Avolio, B., B. Bernard, and D. Jung. "Re-examining the Components of Transformational and Transactional Leadership Using the Multifactor Leadership Questionnaire." *Journal of Occupational and Organizational Psychology* 72 (1999): 441–62.

Avolio, B., J. Lawler, and F. Walumbwa. "Leadership, individual differences, and work-related attitudes: A cross-culture investigation." *Applied Psychology: An International Review* 56, no. 2 (2007): 212–30.

Baban, A., and L. Ratiu. "Executive coaching as a change process: An analysis of the readiness for coaching." *Cognition, Brain, Behavior: An Interdisciplinary Journal* 16, no. 1 (2012): 139–64.

Barge, J., and D. Schlueter. "Leadership as Organizing: A Critique of Leadership Instruments." *Management Communication Quarterly* 4 (1991): 541–70.

Barling, J., M. Beauchamp, L. Masse, K. Morton, R. Rhodes, and B. Zumbo. "Extending Transformational Leadership Theory to Parenting and Adolescent Health Behaviors: An integrative and theoretical review." *Health Psychology Review* 4, no. 2 (2010): 128–57.

Bateman, T. S., and C. P. Zeithaml. *Management: Function and Strategy.* K. Strand, J. Roberts, and L. Ruberstein, eds. Homewood, IL: Irwin, 1993.

Belschak, F., N. Deanne, and D. Hartog. "When does transformational leadership enhance employee proactive behavior? The role of autonomy and role breadth self-efficacy." *Journal of Applied Psychology* 97, no. 1 (2012): 194–202.

Bennett, T. "A Study of the Management Leadership Style Preferred by its Subordinates." *Journal of Organizational Culture* 13, no. 2 (2009): 1–24.

Bolıvar-Cruz, A., I. Galvan-Sanchez, S. A. Gonzalez-Betancor, P. McCauley-Bush, K. Meza, A. Millan, D. Miranda, L. Rabelo, and B. Rodriguez. "Assessing oral presentation skills in Electrical Engineering: Developing a valid and reliable rubric." *International Journal of Electrical Engineering Education* 54, no. 1 (2017): 17–34.

Bonasso, R. E., and A. McKee. "Resonant Leadership: Renewing Yourself and Connecting with Others through Mindfulness, Hope, and Compassion." *Journal of Professional Issues in Engineering Education and Practice* 127, no. 1 (2005): 17–25.

Boyatzis, S. *Engineering Leadership and Integral Philosophy.* Boston: Harvard Business School Press, 2001.

Boyatzis, R. E., D. Goleman, and A. McKee. *Primal Leadership: Realizing the Power of Emotional Intelligence.* Boston: Harvard Business School Press, 2002.

Brackett, M., S. Rivers, and P. Salovey. "Emotional Intelligence: Implications for Personal, Social, Academic, and Workplace Success." *Social and Personality Psychology Compass* 5, no. 1 (2011): 88–103.

Brazil, D., and J. Farr. "Leadership Skills Development for Engineers." *Engineering Management Journal* 21, no. 1 (2009): 3–8.

Briesch, A., S. Chafouleas, H. Swaminathan, and M. Welsh. "Generalizability theory: A practical guide to study design, implementation, and interpretation." *Journal of School Psychology* 52 (2014): 13–36.

Brooks, A., K. Levine, and R. Muenchen. "Measuring Transformational and Charismatic Leadership: Why isn't charisma measured?" *Communications Monographs* 77, no. 4 (2010): 576–91.

Burton, D. "The most hazardous and dangerous and greatest adventure on which man has ever embarked." *Mechanical Engineering* 131, no. 7 (2009): 28–35.

Carless, S., L. Mann, and A. Wearing. "Leadership, Managerial Performance and 360-Degree Feedback." *Applied Psychology: An International Review* 47, no. 4 (1998): 481–96.

Carlson, J. G. "Recent assessments of the Myers-Briggs type indicator." *Journal of Personality Assessment* 49, no. 4 (1985): 356–64.

Cilliers, F., V. Deventer, and R. Eeden. "Leadership Styles and Associated Personality Traits: Support for the conceptualization of transactional and transformational leadership." *South African Journal of Psychology* 38, no. 2 (2008): 253–57.

Čížek, P. "The Application of Maslow's Hierarchy of Needs to the Entrepreneur's Motivation—The Example from Region Pardubice." *Scientific Papers of the University of Pardubice* 24, no. 2 (2012): 43–50.

Clegg, C., J. Cordery, and T. Wall. "Empowerment, performance, and operational uncertainty: A theoretical integration." *Applied Psychology: An International Review* 51, no. 1 (2002), 146–69.

Collado, J., J. L. Laglera, and J. A. Montes. "Effects of Leadership on Engineers: A Structural Equation Model." *Engineering Management Journal* 25 (2013): 7–16.

Copi, I. M. *Introduction to Logic.* C. Owen, R. Miller, S. Lesan, C. Smith, D. O'Connell, and D. Chodoff, eds. Upper Saddle River, NJ: Pearson Education, Inc., 2002.

Creswell, J. W. *Research Design: Qualitative, Quantitative and Mixed Methods Approaches.* Thousand Oaks, CA: Sage Publications, 2009.

Crispo, A. and Y. Sysinger. "Employee motivation and 360 degree feedback." *Insights to a Changing Worlds Journal* 9 (2012): 1–13.

Crumpton-Young, L., A. Ferreras, M. Fernandez-Monroy, M. Kelarestani, and D. Verano-Tacoronte. "Engineering Leadership Development Programs: A Look at What is Needed and What is Being Done." *Journal of STEM Education* 11, no. 3/4 (2016): 10–21.

Daley, J. History's Greatest Speeches. (J. Daley, M. Waldrep, and J. Kopito, eds.) Mineola: Dover Thrift Editions, 2013.

Dainty, A., J. Kissi, and M. Tuuli. "Examining the Role of Transformational Leadership of Portfolio Managers in Project Performance.: *International Journal of Project Management* 31 (2011): 485–97.

Deanne, N., D. Hartog, and E. Keegan. "Transformational Leadership in a Project Based Environment: A comparative study of the leadership styles of project managers and line managers." *International Journal of Project Management* 22 (2004): 609–17.

De Berg, R., A. Jarzebowski, and J. Palermo. "When feedback is not enough: The impact of regulatory fit on motivation after positive feedback.: *International Coaching Psychology Review* 7, no. 1 (2012): 14–32.

Dubrin, A. *Leadership: Research Findings, Practice, and Skills*. Mason, OH: South-Western, Cengage Learning, 2010.

Elster, C. H. Verbal Advantage. Achievement Dynamics Corporation. 1995.

Ellam-Dyson, V., and S. Palmer. "Leadership coaching? No thanks, I'm not worthy." *The Coaching Psychologist* 7, no. 2 (2011): 108–17.

Elliott, A., and W. Woodward. *Statistical Analysis Quick Reference Guidebook With SPSS Examples*. Thousand Oaks, CA: Sage Publications, 2007.

Ferris, G. R., T. P. Munyon, A. A. Perryman, and J. K. Summers. "Dysfunctional executive behavior: What can organizations do?" *Business Horizons* 53 (2010): 581–90.

Forbes. "Why Engineers Make Great CEOs." May 2014.

French, J. R., and B. Raven. "The Bases of Power and the Power/Interaction Model of Interpersonal Influence." *Analyses of Social Issues and Public Policy*, 8, no. 2, (2008): 1-22.

Funk and Wagnalls. *Young Students Learning Library: Universe—World War I*, Volume 22, pp. 2,802–3. Shelton, CT. 1995.

Grant, R. M. *Contemporary Strategy Analysis: Concepts, Techniques, Applications*, second edition. Cambridge: Blackwell Publishers, 1995.

Harvard Business Review. "The Results Driven Manager: Presentations That Persuade and Motivate." Boston: Harvard Business School Press, 2004.

Harvard Business Review. "The Best-Performing CEOs in the World 2018." Harvard Business Review Staff, 2018.

Harvard Business Review. The Best-Performing CEOs in the World." Harvard Business Review Staff, 2014.

Hathaway, R. "Assumptions Underlying Quantitative and Qualitative Research: Implications for Institutional Research.: *Research in Higher Education* 36, no. 5 (1995): 535–62.

Hosein, Z., and A. Yousefi. "The role of emotional intelligence on workforce agility in the workplace." *International Journal of Psychological Studies* 4, no. 3 (2012): 48–61.

Hu, J., and R. Liden. "Antecedents of team potency and team effectiveness: An examination of goal and process clarity and servant leadership." *Journal of Applied Psychology* 96, no. 4 (2011): 851–62.

Ivancevich, J., and M. Matteson. *Organizational Behavior and Management: Third Edition.* K. Strand, L. Spell, and R. McMullen, eds. Burr, IL: Irwin, 1993.

Johns, H. E., and H. R. Moser. "From trait to transformation: The evolution of leadership theories." *Academic Journal Education* 110, no. 1 (1989): 115–23.

Juran, J. M. *Managerial Breakthrough: The Classical Book on Improving Management Performance.* New York: McGraw-Hill, 1995.

Kark, R., and T. Yaffe. "Leading by example: The case of leader OCB." *Journal of Applied Psychology* 96, no. 4 (2011): 806–26.

Kets, V. M. F. R., K. Kortov, and E. Florent-Treacy, eds. *Coach and Couch: The Psychology of Making Better Leaders.* Basingstoke, England: Palgrave Macmillan, 2007.

Kouzes, J. M., and B. Z. Posner. *The Leadership Challenge.* San Francisco: Jossey-Bass, 2007.

Korner, H., and H. Nordvik. "Personality traits in leadership behavior." *Scandinavian Journal of Psychology* 45 (2004): 49–54.

Kussrow, P. "Brain-based leadership.: *Contemporary Education* 72, no. 2 (2009): 10–15.

Labarre, R. "Themes of State Power: People." *Baltic Defense Review* 12, no. 2 (2004): 30–36.

Lam, S., A. Peng, and J. Schaubroeck. "Cognition-based and affect-based trust as mediators of leader behavior influences on team performance." *Journal of Applied Psychology* 96, no. 4 (2011): 863–71.

Lawrence, P. "What leaders need to know about human evolution and decision making." *Leader to Leader* 60 (2011): 12–16.

Leedy, P., and J. Ormond. *Practical Research Planning and Design.* Upper Saddle River, NJ: Pearson Education, Inc., 2013.

Levine, H., and D. A. Raynor. "Associations between the five-factor model of personality and health behaviors among college students." *Journal of American College Health* 58, no. 1 (2006): 73–82.

Liu, J., S. Siu, and K. Shi. "Transformational leadership and employee well-being: The mediating role of trust in the leader and self-efficacy." *Applied Psychology: An International Review* 59, no. 3 (2010): 454–79.

Lo, May-Chiun, H. W. Min, and T. Ramayah. "Leadership styles and organizational commitment: A test on Malaysia manufacturing industry." *African Journal of Marketing Management* 1, no. 6 (2009): 133–39.

Lynam, D., and J. D. Miller. "Structural models of personality and their relation to antisocial behavior: A meta-analytic review." *Academic Journal—Criminology* 39, no. 4 (2001): 765–98.

McElhaney, J. "Legalese Just Gets in the way of Effective Communication." *Science & Engineering Ethics* 96, no. 1 (2010): 22–23.

McCuen, R. "A Course on Engineering Leadership." *Journal of Professional Issues in Engineering Education and Practice* 125, no. 3 (1999): 79–82.

Muller, R., Turner, H. Rodney. "Matching the Project Manager's Leadership Style to Project Type." *International Journal of Project Management* 25 (2007): 21–32.

Murphy, H., and C. Peck. *Effective Business Communications.* New York: McGraw-Hill, 1972.

Neuhauser, C. "Project Manager Leadership Behaviors and Frequency of Use by Female Project Managers." *Project Management Journal* 38, no. 4 (2007): 47–54.

Noonan, D. "Preparing and conducting interviews to collect data." *Nurse Researcher* 20, no. 5 (2013): 28–32.

Northouse, P. G. *Leadership: Theory and Practice: Sixth Edition.* L. Shaw, P. Quinlin, M. Stanley, M. White, N. Vail, E. Garner, and M. Masson, eds. Thousand Oaks, CA: Sage Publications, 2013.

Northouse, P. G. *Leadership: Theory and Practice: Fifth Edition*. L. Shaw, P. Quinlin, M. Stanley, M. White, N. Vail, E. Garner, K. Ehrmann, and M. Masson, eds. Thousand Oaks, CA: Sage Publications, 2013.

Plinio, A. J. "Ethics and Leadership." *International Journal of Disclosure and Governance* (2009): 277–83.

Prabhakar, G. "An Empirical Study Reflecting the Importance of Transformational Leadership on Project Success across Twenty-Eight Nations." *Project Management Institute Journal* 36, no. 4 (2005): 53–60.

Reynolds, G. *Presentationzen: Simple Ideas on Presentation Design and Delivery*. M. Nolan, M. Justak, and H. Sala, eds. Berkeley: New Riders, 2008.

Rock, D., and J. Schwartz. "The neuroscience of leadership." *Reclaiming Childhood and Youth* 16, no. 3 (2007): 10–17.

Romanowski, C., and P. Sageev. "A Message from Recent Engineering Graduates in the Workplace: Results of a Survey on Technical Communication Skills." *Journal of Engineering Education* ASEE (2001): 685–93.

Rylander, P. "Coaches' Power Bases of Power: Developing Some Initial Knowledge of Athletes' Compliance with Coaches in Team Sports." *Journal of Applied Sports Psychology* 27 (2015): 110–21.

Shipside, S. *Effective Communication: Get Your Message Across and Learn How to Listen*. F. Biggs, T. Jeavons, J, Lanaway, M. Hemsley, S. Tuite, T. Broder, M. Parrish, and A. Haywood, eds. New York: Dorling Kindersley, 2007.

Sibel, O., S. G. Olga, J. Alabart, and M. Medir. "Assessment of Engineering Students' Leadership Competencies." *Leadership and Management in Engineering* (2013): 65–75.

Spieth, P., A. Tyssen, and A. Wald. "The Challenge of Transactional and Transformational Leadership in Projects." *International Journal of Project Management* (2013): 4–16.

Srivastava, S. "The five-factor model describes the structure of social perceptions." *Psychology Inquiry* 21 (2010): 69–75.

Steensma, H., and B. van Knippenberg. "Future interaction expectation and the use of soft and hard influence tactics." *Applied Psychology: An International Review* 52, no. 1 (2003): 55–67.

Suleyman, Y. "Relationship between leader-member exchange and burnout in professional football players." *Journal of Sports Sciences* 29, no. 14 (2011): 1,493–502.

Tapanainen, T., and A. Ryoma. "The Applicability of Transformational Leadership to Short-term Projects." *Turku School of Economics Finland* 2 (1999): 332–38.

Vroom, V. "Loose-tight leadership: What is the question?" *Applied Psychology: An International Review* 46, no. 4 (1997): 422–27.

Weingardy, R. "Engineers and Leadership: How to Move up the Economic Food Chain." *J. Proff Issues Eng. Edu. Prac* 120, no. 1 (1994): 50–57.

Welch, Jack. *Straight from the Gut*. J. Byrne, ed. New York: Warner Business Books, 2001.

APPENDIX A: DETAILED LITERATURE GAP ANALYSIS

Article No.	Article	Assessment of PM TL	Assessment of PM TZ	Assessment of TL		Assessment of TZ		Engineering Degree
				Certified PM's	Non-Certified PM's	Certified PM's	Non-Certified PM's	
1	TL in a project-based environment: a comparative study of the leadership styles of project managers and line managers.	X	contingent reward behavior	GAP	GAP	GAP	GAP	GAP
2	Project Manager Leadership Role in Improving Project Performance	(reference to Hartog & Keegan's TL study)	GAP	GAP	GAP	GAP	GAP	GAP
3	The Applicability of TL to Short-term Projects.	in projects not individuals	GAP	GAP	GAP	GAP	GAP	GAP
4	Project Manager Leadership Behaviors and Frequency of Use by Female Project Managers	Females	Females	GAP	GAP	GAP	GAP	GAP
5	Leadership competency profiles of successful project managers.	(reference to Hartog & Keegan's TL study)	In projects	GAP	GAP	GAP	GAP	GAP
6	Understanding the Role of Vision in	Project Vision	GAP	GAP	GAP	GAP	GAP	GAP

		Project Success							
7	An Empirical Study Reflecting the importance of TL on Project Success Across Twenty-Eight Nations	Project Success	GAP	GAP	GAP	GAP	GAP	GAP	GAP
8	Examining the role of TL of portfolio managers in project performance	Portfolio Managers	GAP	GAP	GAP	GAP	GAP	GAP	GAP
9	The challenge of TZ and TL in projects.	X	X	GAP	GAP	GAP	GAP	GAP	GAP
10	The association among project manager's leadership style, teamwork and project success.	X	X	GAP	GAP	GAP	GAP	GAP	GAP
11	Matching the project manager's leadership style to project type.	(reference to Hartog & Keegan's TL study)	for engineering projects	GAP	GAP	GAP	GAP	GAP	GAP
12	A Study of the Management Leadership style preferred by IT subordinates.	studied IT managers and subordinates	GAP	GAP	GAP	GAP	GAP	GAP	GAP
13	The Project Manager's Leadership Style As a success factor on Projects: A literature Review	(reference to Hartog & Keegan's TL study)	GAP	GAP	GAP	GAP	GAP	GAP	GAP

APPENDIX B: MULTIFACTOR LEADERSHIP QUESTIONNAIRE

RESULTING FROM PUBLISHING copyright constraints, only five each of the forty-five rater questions/items are shown on the corresponding rater form.

Multifactor Leadership Questionnaire
Rater Form

Name of Leader: _____ Date: _____

Organization ID #: _____ Leader ID #: _____

This questionnaire is used to describe the leadership style of the above-mentioned individual as you perceive it. Answer all items on this answer sheet. If an item is irrelevant, or if you are unsure or do not know the answer, leave the answer blank. Please answer this questionnaire anonymously.

Important (necessary for processing): Which best describes you?

___ I am at a higher organizational level than the person I am rating.
___ The person I am rating is at my organizational level.
___ I am at a lower organizational level than the person I am rating.
___ Other than the above.

Forty-five descriptive statements are listed on the following pages. Judge how frequently each statement fits the person you are describing. Use the following rating scale:

Not at all	Once in a while	Sometimes	Fairly often	Frequently, if not always
0	1	2	3	4

The Person I Am Rating...

1. Provides me with assistance in exchange for my efforts 0 1 2 3 4
2. *Re-examines critical assumptions to question whether they are appropriate 0 1 2 3 4
3. Fails to interfere until problems become serious 0 1 2 3 4
4. Focuses attention on irregularities, mistakes, exceptions, and deviations from standards 0 1 2 3 4
5. Avoids getting involved when important issues arise 0 1 2 3 4

Reproduction by special permission of the Publisher, Mind Garden, Inc., www.mindgarden.com from the Multifactor Leadership Questionnaire by Bernard M. Bass & Bruce Avolio Copyright 1995 by Bernard Bass and Bruce J. Avolio

APPENDIX C: SAMPLE MLQ5X SCORING KEY

RESULTING FROM PUBLISHING copyright constraints, only five each of the forty-five rater questions/items are shown on the corresponding scoring key.

MLQ Multifactor Leadership Questionnaire
Scoring Key (5x) Short

My Name: _____ Date: _____

Organization ID #: _____ Leader ID #: _____

Scoring: The MLQ scale scores are average scores for the items on the scale. The score can be derived by summing the items and dividing by the number of items that make up the scale. If an item is left blank, divide the total for that scale by the number of items answered. All of the leadership style scales have four items, Extra Effort has three items, Effectiveness has four items, and Satisfaction has two items.

Not at all	Once in a while	Sometimes	Fairly often	Frequently, if not always
0	1	2	3	4

*Idealized Influence (Attributed) total/4 =
*Idealized Influence (Behavior) total/4 =
*Inspirational Motivation total/4 =
*Intellectual Stimulation total/4 =
*Individual Consideration total/4 =
\# Contingent Reward total/4 =

\# Management-by-Exception (Active) total/4 =
+Management-by-Exception (Passive) total/4 =
+Laissez-faire Leadership total/4 =
Extra Effort total/3 =
Effectiveness total/4 =
Satisfaction total/2 =

1. Contingent Reward..0 1 2 3 4
2. Intellectual Stimulation..0 1 2 3 4
3. Management-by-Exception (Passive)...........................0 1 2 3 4
4. Management-by-Exception (Active).............................0 1 2 3 4
5. Laissez-faire Leadership...0 1 2 3 4

Reproduction by special permission of the Publisher, Mind Garden, Inc., www.mindgarden.com from the Multifactor Leadership Questionnaire by Bernard M. Bass & Bruce Avolio Copyright 1995 by Bernard Bass and Bruce J. Avolio

APPENDIX D: NORMATIVE TABLES

Percentiles for Individual Scores Based on Lower Level Ratings (US)

%tile	II(A) 12,118	II(B) 12,118	IM 12,118	IS 12,118	IC 12,118 MLQ Scores	CR 12,118	MBEA 12,118	MBEP 12,118	LF 12,118	EE 12,118	EFF 12,118 Outcomes	SAT 12,118	%tile
5	1.25	1.25	1.5	1.5	1	1.29	0.25	0	0	1	1.5	1	5
10	1.75	1.75	2	1.75	1.5	1.75	0.5	0	0	1.33	2	2	10
20	2.25	2.21	2.25	2.25	2	2.25	0.75	0.25	0	2	2	2.5	20
30	2.5	2.5	2.75	2.5	2.5	2.5	1.11	0.5	0.25	2.33	2.5	3	30
40	2.75	2.54	3	2.75	2.75	2.75	1.37	0.75	0.25	2.67	2.75	3	40
50	3	2.75	3	2.75	3	3	1.62	1	0.5	3	3	3.5	50
60	3.25	3	3.25	3	3.17	3.13	1.87	1	0.75	3	3.25	3.5	60
70	3.5	3.25	3.5	3.25	3.25	3.25	2.25	1.25	0.93	3.33	3.5	3.67	70
80	3.75	3.46	3.75	3.5	3.5	3.5	2.5	1.7	1.25	3.67	3.52	4	80
90	4	3.75	4	3.75	3.75	3.75	3	2	1.75	4	4	4	90
95	4	3.75	4	4	4	4	3.25	2.5	2	4	4	4	95

Percentiles for Individual Scores Based on Same Level Ratings (US)

%tile	II(A) 5,185	II(B) 5,185	IM 5,185	IS 5,185	IC 5,185 MLQ Scores	CR 5,185	MBEA 5,185	MBEP 5,185	LF 5,185	EE 5,185	EFF 5,185 Outcomes	SAT 5,185	%tile
5	1.5	1.5	1.5	1.5	1.5	1.75	0.25	0	0	0	1.75	1.5	5
10	2	1.75	1.75	1.75	1.75	2	0.5	0.11	0	1.66	2	2	10
20	2.25	2.25	2.25	2.25	2.25	2.37	1	0.35	0	2	2.5	2.5	20
30	2.67	2.5	2.5	2.5	2.5	2.6	1.25	0.5	0.25	2.23	2.75	2.91	30
40	2.75	2.75	2.75	2.75	2.75	2.75	1.5	0.75	0.25	2.67	3	3	40
50	3	2.75	3	2.75	3	3.06	1.75	1	0.5	2.73	3.03	3.08	50
60	3.25	3	3	3	3	3.25	2	1.04	0.75	3	3.25	3.5	60
70	3.5	3.25	3.25	3.25	3.25	3.25	2.25	1.25	1	3.33	3.5	3.5	70
80	3.5	3.28	3.28	3.34	3.34	3.5	2.5	1.5	1.17	3.34	3.75	4	80
90	3.75	3.75	3.75	3.75	3.75	3.75	2.87	2	1.5	3.67	4	4	90
95	4	3.75	4	4	4	4.77	3.25	2.5	2.5	4	4	4	95

Percentiles for Individual Scores Based on Higher Level Ratings (US)

%tile	II(A) 4,268	II(B) 4,268	IM 4,268	IS 4,268	IC 4,268 MLQ Scores	CR 4,268	MBEA 4,268	MBEP 4,268	LF 4,268	EE 4,268	EFF 4,268 Outcomes	SAT 4,268	%tile
5	1.75	1.75	1.5	1.5	1.5	1.75	0.25	0	0	1.33	1.75	1.5	5
10	2	2	1.75	1.75	2	2	0.5	0.25	0	1.67	2	2	10
20	2.5	2.5	2.25	2.18	2.25	2.43	0.95	0.35	0	2	2.5	2.5	20
30	2.75	2.75	2.5	2.41	2.5	2.62	1.25	0.5	0.25	2.33	3	3	30
40	2.95	2.95	2.75	2.5	2.75	2.75	1.5	0.75	0.25	2.67	3.04	3	40
50	3	3	2.9	2.75	2.97	3	1.7	1	0.5	2.74	3.25	3.08	50
60	3.25	3.25	3	3	3	3	1.95	1.03	0.75	2.82	3.5	3.5	60
70	3.5	3.5	3.25	3	3.25	3.25	2.21	1.25	0.92	3	3.5	3.5	70
80	3.5	3.5	3.5	3.25	3.96	3.47	2.5	1.5	1.17	3.33	3.75	3.5	80
90	3.75	3.75	3.75	3.5	3.67	3.62	2.88	2	1.5	3.67	4	4	90
95	4	4	4	3.75	3.75	3.75	3.25	2.5	2	4	4	4	95

Reproduction by special permission of the Publisher, Mind Garden, Inc., www.mindgarden.com from the Multifactor Leadership Questionnaire by Bernard M. Bass & Bruce Avolio Copyright 1995 by Bernard Bass and Bruce J. Avolio

APPENDIX E: FULL-RANGE LEADERSHIP THEORY

TRANSFORMATIONAL LEADERSHIP IS a process of influencing in which leaders change their associates' awareness of what is important and move them to see themselves and the opportunities and challenges of their environment in a new way. Transformational leaders are proactive. They seek to optimize individual, group, and organizational development and innovation, not just achieve performance at expectations. They convince their associates to strive for higher levels of potential as well as higher levels of moral and ethical standards.

A. Idealized Influence (Attributes and Behaviors)

These leaders are admired, respected, and trusted. Followers identify with and want to emulate their leaders. Among the things leaders do to earn credit with followers is considering followers' needs over their own needs. Leaders shares risks with followers and are consistent in conduct with underlying ethics, principles, and values.

B. Inspirational Motivation (IM)

These leaders behave in ways that motivate those around them by providing meaning and challenge to their followers' work. Individual and team spirit is aroused. Enthusiasm and optimism are displayed. The leader encourages followers to envision attractive future states that they can ultimately envision for themselves.

C. Intellectual Stimulation (IS)

These leaders stimulate their followers' effort to be innovative and creative by questioning assumptions, reframing problems, and approaching old situations in new ways. There is no ridicule or public criticism of individual members' mistakes. New ideas and creative solutions to problems are solicited from followers, who are included in the process of addressing problems and finding solutions.

D. Individual Consideration (IC)

These leaders pay attention to each individual's need for achievement and growth by acting as coaches or mentors. Followers are developed to successively higher levels of potential. New learning opportunities are created along with a climate in which to grow. Individual differences in terms of needs and desires are recognized.

Transactional Leadership

Transactional leaders display behaviors associated with constructive and corrective transactions. The constructive style is labeled contingent-reward and the corrective style is labeled management-by-exception. Transactional leadership defines expectations and promotes performance to achieve these levels. Contingent reward and management-by-exception are two core behaviors associated with management functions in organizations. Full-range leaders do this and more.

A. Contingent Reward (CR)

Transactional contingent-reward leadership clarifies expectation and offers recognition when goals are achieved. The clarification of goals and objectives and the providing of recognition once goals are achieved should result in individuals and groups achieving expected levels of performance.

Reproduction by special permission of the Publisher, Mind Garden, Inc., www.mindgarden.com from the Multifactor Leadership Questionnaire by Bernard M. Bass & Bruce Avolio Copyright 1995 by Bernard Bass and Bruce J. Avolio

B. Management-by-Exception: Active (MBEA)

The leader specifies the standards for compliance as well as what constitutes ineffective performance and may punish followers for being out of compliance with those standards. This style of leadership implies closely monitoring of deviances, mistakes, and errors and then taking corrective action as quickly as possible when they occur.

C. Passive-Avoidant Behavior

Another form of management-by-exception leadership is more passive and reactive. It does not respond to situations and problems systematically. Passive leaders avoid specifying agreements, clarifying expectations, and providing goals and standards to be achieved by followers. This style has a negative effect on desired outcomes—the opposite of what is intended by the leader-manager. In this regard, it is similar to laissez-faire styles of no leadership. Both types of behavior have negative impacts on followers and associates. Accordingly, both styles can be grouped together as passive-avoidant leadership.

Management-by-Exception: Passive (MBEP)

Laissez-Faire (LF)

Outcomes of Leadership

Transformational and transactional leadership are both related to the success of the group. Success is measured with the MLQ by how often the raters perceive their leader to be motivating, how effective raters perceive their leader to be at interacting at different levels of the

Reproduction by special permission of the Publisher, Mind Garden, Inc., www.mindgarden.com from the Multifactor Leadership Questionnaire by Bernard M. Bass & Bruce Avolio Copyright 1995 by Bernard Bass and Bruce J. Avolio

organization, and how satisfied raters are with their leader's methods of working with others.

Reproduction by special permission of the Publisher, Mind Garden, Inc., www.mindgarden.com from the Multifactor Leadership Questionnaire by Bernard M. Bass & Bruce Avolio Copyright © 1995 by Bernard Bass and Bruce J. Avolio

APPENDIX F: SURVEY PROCESS

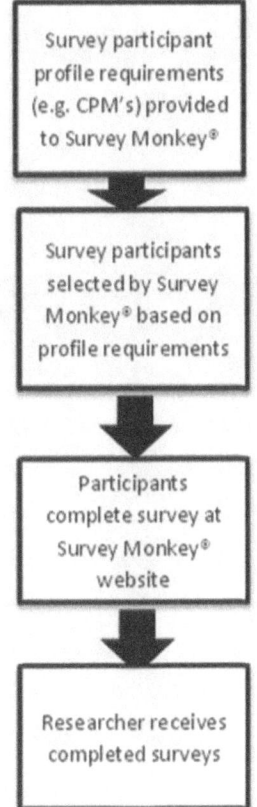

APPENDIX G: NONPARAMETRIC TESTING

EMPLOYING NONPARAMETRIC TESTING, the intent of this appendix is to confirm findings resulting from the use of parametric testing of key aspects of the hypothesis H1, which says that there is no statistically significant difference between leadership styles of managers (CPMs, non-CPM managers, or the integrated manager group) with engineering degrees in comparison to the same without engineering degrees.

The first test considers whether or not there is any statistically significant difference in TL as measured by perceived style scores. The Mann-Whitney U test will be employed using SPSS. Upon running this test, the summary information is indicated in table 1. The Mann-Whitney Test in SPSS automatically restates the hypothesis as indicated in the "null hypothesis" column. The "Sig" or "P value" is also given, which, based on its level, corresponds to the "Decision" column output. In this case, given that p equals 0.397, H1 cannot be rejected, which is consistent with the parametric test findings using the two sample tests.

Table 1: CPM Summary Information

Hypothesis Test Summary

	Null Hypothesis	Test	Sig.	Decision
1	The distribution of LdrStyle is the same across categories of CPMs.	Independent-Samples Mann-Whitney U Test	.397	Retain the null hypothesis.

Asymptotic significances are displayed. The significance level is .05.

Figure 2 presents the distributions that must also be considered with the Mann-Whitney U application. SPSS uses the "population pyramid" method for presenting these distributions. And as exhibited in composite figure 2, the distributions for CPM populations appear relatively similar. (Note that whether the reader views the distributions as similar or dissimilar does not change the statistical result.) Table 3 reflects the medians associated with the CPM populations.

Figure 2: Independent Samples Mann-Whitney U Test (CPMs)

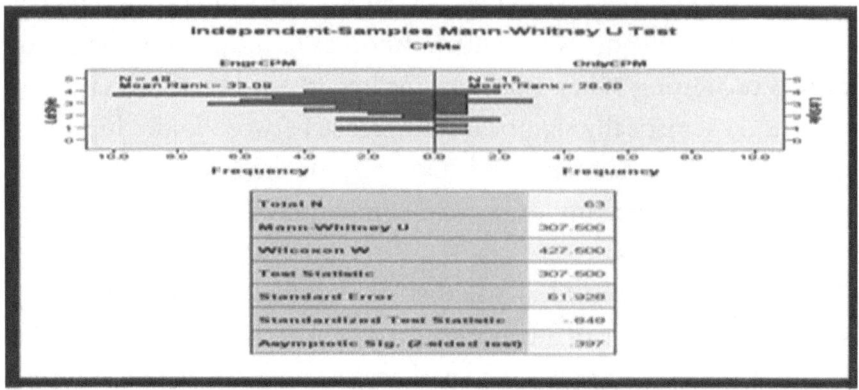

Table 3: CPM Medians

CPMs	LdrStyle
EngrCPM	3.1500
OnlyCPM	2.9000
Total	3.1000

Based on the previously outlined data, the aggregated results regarding the CPM populations with and without engineering degrees are expressed as follows: A Mann-Whitney U test was run to determine if there were differences in LeaderStyle scores between EngrCPMs and OnlyCPMs. Distributions of the LeaderStyle scores for EngrCPMs and OnlyCPMs were similar as assessed by visual inspection. LeaderStyle score was not statistically significantly different between EngrCPMs (Mdn: 3.15) and OnlyCPMs (Mdn: 2.90), U = 307.5, z = -0.848, p = .397.

The Mann-Whitney U test was also implemented with the manager and integrated populations. As was the case previously, both tests confirmed the results achieved employing the parametric t-test.

With managers, a Mann-Whitney U Test was run to determine if there were differences in LeaderStyle scores between EngrMgrs and OnlyMgrs. Distributions of the LeaderStyle scores for EngrMgrs and OnlyMgrs were similar, as assessed by visual inspection. LeaderStyle score was statistically significantly higher in EngrMgrs (Mdn = 2.73) than in OnlyMgrs (Mdn = 2.08), U = 3,588, z = -3.161p = .002.

With integrated populations, a Mann-Whitney U test was run to determine if there were differences in LeaderStyle scores between EngrINTGR and OnlyINTGR. Distributions of the LeaderStyle scores for EngrINTGR and OnlyINTGR were similar, as assessed by visual inspection. Median LeaderStyle scores were statistically significantly higher in EngrINTGR (2.85) than in OnlyINTGR (2.20), U = 6250.5.5, z = -4.8109, p = .0005.

Further parametric testing of the H1 hypothesis included analysis of the five TL constituents aimed at determining whether or not statistically significant differences existed. ANOVA was employed for this parametric testing. The nonparametric test employed in this section is the Kruskal-Wallis H test. Table 4 provides summary information associated with the Kruskal-Wallis H test results.

Table 4: TL Attribute Summary Data

Hypothesis Test Summary

	Null Hypothesis	Test	Sig.	Decision
1	The distribution of Attribute Value is the same across categories of TL Atribute.	Independent-Samples Kruskal-Wallis Test	.313	Retain the null hypothesis.

Asymptotic significances are displayed. The significance level is .05.

In addition to table 4, composite figure 20 presents the distributions that must also be considered as well as other necessary statistical data after the implementation of the Kruskal-Wallis H test application.

Unlike the "population pyramid" method mentioned previously, SPSS uses *box plots* for presenting the Kruskal-Wallis H test distributions. As seen in composite figure 20, the distributions for perceived TL constituent scores appear relatively similar. As previously discussed, the five constituent categories tested are idealized attributes (IA), idealized behavior (IB), inspirational motivation (IM), intellectual stimulation (IS), and individualized consideration (IC).

Figure 5: Distribution Data

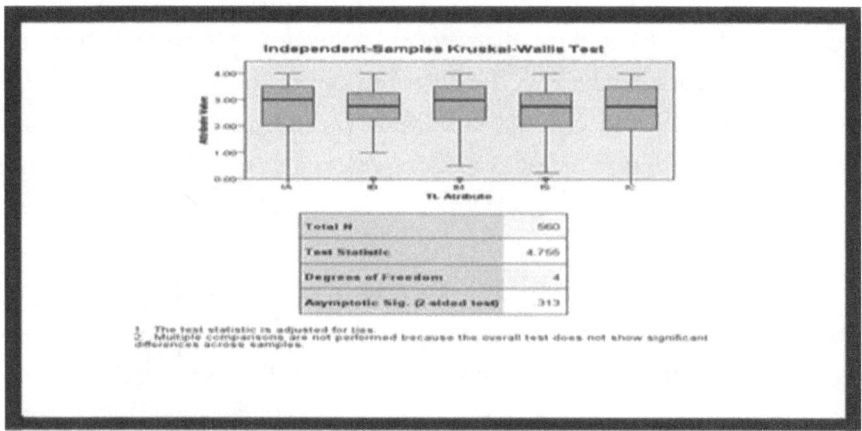

Table 6 reflects the medians associated with the perceived TL constituent levels.

Table 6: Median Levels for TL Attributes

Attribute Value

TL Atribute	N	Median
IA	112	3
IB	112	2.75
IM	112	3
IS	112	2.75
IC	112	2.75
Total	560	3

Referring to composite figure 20, note that the "Test Statistics" row provides the value of the H statistic (4.755), which approximately follows a χ2-distribution (CHI Square) with k–1 degrees of freedom.

Considering the resulting test data in aggregate, the following statements can be made regarding TL attributes:

A Kruskal-Wallis H test was run to determine if there were differences in perceived LeaderStyle score between five groups of scores with different leader style measures—IA, IB, IM, IS, and IC. Distributions of LeaderStyle scores were similar for all groups as assessed by visual inspection of a box plot. Median LeaderStyle scores were equal at (Mdn = 2.75), for IB, IS and IC, and likewise for IM & IA (Mdn = 3.0) but the differences were not statistically significant, $\chi2(4) = 4.755$, $p = .313$.

The previously outlined constituent test results confirm the findings from the previously run parametric testing discussed in the results section of this writing.

Summary

Results from the parametric (two sample independent t-tests) aimed at determining whether or not a statistically significant difference in means occurred between each of the CPM, manager, and integrated populations (with and without engineering degrees) were confirmed employing the nonparametric Mann-Whitney U test.

Likewise, the parametric ANOVA test, which was implemented to identify statistically significant differences in the five TL constituents for the integrated population (with engineering degrees), was also confirmed employing the Kruskal-Wallis H test.

Thus, despite the lack of conformance to some of the assumptions (e.g., outliers, normality, etc.) evident prior to implementing the parametric testing, the author's position of deferring to previously cited comments offered by Elliot and Woodward (2007) remains unchanged and further buttressed by the nonparametric testing results provided in this appendix.

INDEX

A

All the Way to the Top: A Practical Guide for Corporate and Business Leadership (Calloway), 113, 126, 160, 168, 186
ANOVA (analysis of variance), 57, 81, 85–87, 90, 103, 109, 213
Anthony, Susan B., 163–64
arguments, 174–76
  deductive, 175–76
  inductive, 175
Armstrong, A., 73, 151
Avolio, Bruce J., 31, 49, 52–53, 61, 71–74, 83, 104–5, 112, 199, 201, 203, 206–8

B

Barge, 12, 73–74
Barra, Mary, 3
Bass, Bernard M., 31–32, 52, 71, 73–74, 105, 112
Bennet, T., 66, 105, 217
Bethone, Kenneth, 113
Bezos, Jeff, 3
body language, 117, 119–20
Bolivar-Cruz, A., 160, 217
Bonasso, R. E., 4, 6, 51, 106, 145, 217
Boyatzis, S., 7, 21, 23–24, 26–27
Brilliant, Josef
  big break, 140
  feedback, 142
  performance evaluation, 115
  power in action, vi, 137
  role, 148

Burns, J. M., 7, 9, 30, 217

C

coaching, 18–20
  executive, 18, 20
communication channels, 116, 217
  nonverbal, 117, 119, 173
  spoken, 117
  written, 117
communications, 34, 115–17, 121–26, 128, 132, 160, 178–79
Cook, Tim, 3
CR (contingent reward), 53, 72, 74, 105–6, 143, 206
credibility, leadership, 28, 30, 32
Creswell, J. W., 68–69, 71, 73, 217
Cruise, Tom, 173
Crumpton-Young, L., 6

D

Dana (Josef's team member), 151, 155
Daniela (director), 114–15, 126, 137, 140
dashboard, 130–34
data analysis, 79, 149, 183
Douglas (Josef's team member), 151, 155–56
Dubrin, A., 12, 173, 177

E

Easy, Earl, 113
EI (emotional intelligence), domains of, 21–22, 24–25, 35, 183
  relationship management, 22–23, 183, 218

self-awareness, 22–23, 183
self-management, 22–23, 183, 218
social awareness, 22–23, 183, 218
empathy, 23–24
engineering education, 53–56, 67, 69, 76, 79, 104, 183
engineers, 1–2, 4, 6, 43, 51, 55, 116–17, 152, 160, 185

F

feedback, 16–17, 19, 24, 26
FFM (Five-Factor Model), 37–39, 70, 183
firefighting, 27, 143
Fiske, D. W., 38
5CLIM (5C Leadership Improvement Model), 127, 184
5Ps, 160, 164, 172, 174, 176
  perception, 172–74, 178
  persuasion, 174, 176, 178, 181, 185
  power in action, 176–79
  preparation, 164–65, 174, 178–79, 185
  purpose, 164, 166, 174, 178, 185
followers, 8–9, 29, 32, 35, 51, 145, 205–7
FRLT (full-range leadership theory), 71, 183, 205

G

Gandhi, Mohandas Karamchand, 162–63
GELI (global executive leadership inventory), 17–18, 218
generalizability, vi, 108, 110, 184, 218
Gerald (Josef's team member), 151, 157–60, 185
gestures, 117, 120–21
Goode, Johnny, 113
Gorbachev, Mikhail, 176
Grant, R., 130

H

Harold (Josef's team member), 151, 153–54
Hathaway, R., 69
HBR (Harvard Business Review), 1, 161
Hours, Kimberly, 113
hurricane Katrina, 36–37

I

IC (individualized consideration), 49, 51, 74, 150, 184, 206, 214
idealized influence, 29, 49, 51, 66, 72, 205, 218
IM (inspirational motivation), 49, 52, 67, 72, 87, 89, 103, 144, 205, 214
instrument
  reliability of, 70–71, 73–74, 218
  validity of, 70–73, 175
iPO model, 148–49, 184
IS (intellectual stimulation), 31–32, 49, 67, 72, 74, 83, 144, 154, 205, 214, 218

J

Jerry Maguire, 172–73, 218

K

Karla (Josef's team member), 151–53, 218
Kennedy, John F., 9, 169, 218
Kilmann, ralph, 155
Kussrow, P., 34

L

Labarre, Frederic, 48
Lawrence, P., 35–36
LDPs (leadership development programs), 55
leaders, 3, 6–8, 10–12, 15–16, 18, 21–25, 27, 29, 31, 33, 35–37, 42, 49,

51, 62, 67, 96, 105, 112, 145, 148, 151, 154–55, 177, 183, 185
  effective, 34, 53, 105, 150, 156, 173
  transactional, 49, 143, 206
  transformational, 29–32, 49, 51, 65, 72, 145, 205
leadership, 4, 6–7, 9, 12–13, 15, 21, 25, 28, 32, 40, 48, 52, 61, 63, 150, 207
  effective, 20, 31, 37, 53, 61, 105
  engineering, 2, 6–7, 182
  executive, 21
  exemplary, 29
  resonant, 25
  styles of, 54–55, 67, 69, 76
  transactional, 183
LED (light emitting diode), 4, 126
LF (laissez-faire), 72, 143, 207
Lincoln, Abraham, 161, 163

M

managers, 18, 48–49, 56–57, 75, 77, 82, 100, 104, 141, 211, 213, 215
  CPMs (certified project managers), 54–56, 62–63, 77, 83, 212
  non-CPMs, 55–56, 63
MBEA (Management-by-Exception: Active), 53, 72, 105, 143, 207
MBEP (Management-by-Exception: Passive), 72, 143, 207
MBTI (Myers-Briggs Type indicator), 38
McCuen, R., 6, 54, 219
Mehrabian, Albert, 119
Merriam-Webster, 7, 40–41, 117, 127, 161, 164, 173–74
mind garden, 57, 95, 98
MLQ (Multifactor Leadership Questionnaire), 54, 57, 63, 69, 71–74, 81, 85, 95, 141, 144, 207
motivation, 25–26, 29–30, 37, 56, 65

N

Neuhauser, C., 64–65
Nixon, Richard M., 163
nonparametric tests
  Kruskal-Wallis, 108, 213–15, 219
  Mann-Whitney U, 108, 211–13
Noonan, D., 68
norm table, 95–96, 98–102
Northouse, P. G., 7, 10–12, 22, 40, 150
Nuttawuth, M., 73

O

ODU IRB (Old Dominion University Internal Review Board), 57–58
OEM (original equipment manufacturers), 138, 157
oration, 161, 178
organization, 1, 6–7, 18, 23–24, 29–32, 35, 42, 46, 72, 114, 128, 131, 148–49, 151, 178, 206, 208
OSHA (Occupational Health and Safety Organization), 134

P

PA (passive avoidant), 72, 79, 81–82, 91–95, 102–3, 107, 207
Page, Larry, 3
peers, 24, 44, 48
personality traits, five big, 38–39
  agreeableness, 38, 219
  conscientiousness, 38
  extraversion, 38, 219
  neuroticism, 38, 219
  openness, 38
PID (proportional-integral-derivative), vii, 138
pitch, 118–19
population, integrated, 75–78, 83–84, 88–89, 101–2, 104–5, 107, 109, 144, 213, 215

Powell, Colin, 164
power, 40–41, 43–44, 48–49
power bases, five, 41, 49–53, 183
   coercive, 41, 44–45, 47, 50
   expert, 41–43, 47, 51, 53, 177
   legitimate, 41, 43–44, 50
   referent, 41, 45–46, 51–52, 177, 179
   reward, 41, 45, 50
power stress, 26–27
presentation, 161, 164–66, 168, 173, 177–78, 180
   formal, 160–61
psychology, 32, 37, 183
purpose, 164

R

reagan, ronald, 176
relationships, in-group, 12
research
   current, 69–70, 73, 75, 84, 108–10
   first hypotheses, 79, 81, 104, 106–7
   second hypotheses, 76, 85, 106
resonance, 25–26, 33, 128, 183
Reynolds, G., 166
Roosevelt, Franklin, 162–63
Rudolph the Red-Nosed Reindeer, 47

S

Shaw, P., 7
sig value, 86–87, 89, 92, 106
skills
   soft, 2
   technical, 5, 54, 116
sMarT mnemonic, 151–52
speech, 161, 164, 178–79
speed, 118
Srivastava, S., 38–39
Sterling, Dave, 112
subordinates, 6, 10–12, 24, 97, 104

Summers, Maria, 113, 126
Summit Consumables Incorporated, 110–13, 134, 141, 154, 161
survey Monkey, 57, 68, 76, 220

T

Tawanda (Josef's team member), 151–52, 220
team leader, 14, 16, 113, 220
theoritical leadership models, 10–11, 13, 32, 182
   contingency, 10
   leader-member exchange (LMX), 11–12
   path-goal, 11, 182
   Vroom-Jago, 11, 13, 220
Thomas, Kenneth, 155
360-degree feedback, 16, 19–20, 141
Tillerson, rex, 3
TL (transformational leadership), 28–30, 32, 49–51, 53–56, 61–67, 72–73, 79, 81, 92–96, 99, 102, 104–8, 110, 112, 143–44, 153, 183, 205, 207, 211
tone, 118–19

V

variables
   dependent. *See* leadership styles
   independent. *See* engineering education
VR2 (variable resistor), 4, 126
Vroom, V., 16, 220

W

Weingardt, R., 2, 43, 185–86, 220
William, 151, 154–55, 220

X

XL (transactional leadership), 10, 28–29, 32, 50, 52–53, 56, 61–62, 71, 73, 79, 81–83, 92, 102, 104–5, 143, 183–84, 206–7

Z

Zellweger, Renee, 173

www.ingramcontent.com/pod-product-compliance
Lightning Source LLC
Chambersburg PA
CBHW021359210526
45463CB00001B/155